T0185145

Geometry and Codes

Mathematics and Its Applications (*Soviet Series*)

V. D. Goppa
Computer Center of the Academy of Sciences of the U.S.S.R., Moscow

Geometry and Codes

Translated by N. G. Shartse

KLUWER ACADEMIC PUBLISHERS
DORDRECHT / BOSTON / LONDON

Library of Congress Cataloging in Publication Data

Goppa, V. D.
    Geometry and codes.

    (Mathematics and its applications ; 24. Soviet series)
    Bibliography: p.
    1. Error-correcting codes (Information theory)
2. Geometry. I. Title. II. Series: Mathematics and
its applications (Kluwer Academic Publishers). Soviet
series ; 24.
QA268.G576 1988   005.6'2   88-9384

Published by Kluwer Academic Publishers,
P.O. Box 17, 3300 AA Dordrecht, The Netherlands.

Kluwer Academic Publishers incorporates
the publishing programmes of
D. Reidel, Martinus Nijhoff, Dr W. Junk and MTP Press.

Sold and distributed in the U.S.A. and Canada
by Kluwer Academic Publishers,
101 Philip Drive, Norwell, MA 02061, U.S.A.

In all other countries, sold and distributed
by Kluwer Academic Publishers Group,
P.O. Box 322, 3300 AH Dordrecht, The Netherlands.

02-0391-200 ts

Printed in The Netherlands

SERIES EDITOR'S PREFACE

Approach your problems from the right end
and begin with the answers. Then one day,
perhaps you will find the final question.

'The Hermit Clad in Crane Feathers' in R.
van Gulik's *The Chinese Maze Murders*.

It isn't that they can't see the solution. It is
that they can't see the problem.

G.K. Chesterton. *The Scandal of Father
Brown* 'The point of a Pin'.

Growing specialization and diversification have brought a host of monographs and textbooks on increasingly specialized topics. However, the "tree" of knowledge of mathematics and related fields does not grow only by putting forth new branches. It also happens, quite often in fact, that branches which were thought to be completely disparate are suddenly seen to be related.

Further, the kind and level of sophistication of mathematics applied in various sciences has changed drastically in recent years: measure theory is used (non-trivially) in regional and theoretical economics; algebraic geometry interacts with physics; the Minkowsky lemma, coding theory and the structure of water meet one another in packing and covering theory; quantum fields, crystal defects and mathematical programming profit from homotopy theory; Lie algebras are relevant to filtering; and prediction and electrical engineering can use Stein spaces. And in addition to this there are such new emerging subdisciplines as "experimental mathematics", "CFD", "completely integrable systems", "chaos, synergetics and large-scale order", which are almost impossible to fit into the existing classification schemes. They draw upon widely different sections of mathematics. This programme, Mathematics and Its Applications, is devoted to new emerging (sub)disciplines and to such (new) interrelations as exempla gratia:

- a central concept which plays an important role in several different mathematical and/or scientific specialized areas;
- new applications of the results and ideas from one area of scientific endeavour into another;
- influences which the results, problems and concepts of one field of enquiry have and have had on the development of another.

The Mathematics and Its Applications programme tries to make available a careful selection of books which fit the philosophy outlined above. With such books, which are stimulating rather than definitive, intriguing rather than encyclopaedic, we hope to contribute something towards better communication among the practitioners in diversified fields.

A *code* is a subset of all the words, of a given length, say, which can be written in a given alphabet, say, the alphabet consisting of the symbols '0' and '1'. If all code words are sufficiently different, a certain number of transmission errors can be detected. This leads to such questions as best error correcting codes for a given transmission rate and others. Other considerations involve efficiency of coding and decoding (and error correction) algorithms.

A number of years ago the author of this book discovered that the theory of algebraic curves (and their Jacobians) over finite fields can be used to construct valuable codes. These are now named Goppa codes and the subject has become an additional important chapter in the general and fast growing field of coding theory. This is, so far, the first book on the subject and both because of its importance and because it links algebraic geometry and coding theory it fits perfectly with the general philosophy of this series as expressed above in the general part of this preface. I am greatly pleased to be able to welcome this unique book in this series.

The unreasonable effectiveness of mathematics in science ...

    Eugene Wigner

Well, if you know of a better 'ole, go to it.

    Bruce Bairnsfather

What is now proved was once only imagined.

    William Blake

As long as algebra and geometry proceeded along separate paths, their advance was slow and their applications limited.

But when these sciences joined company they drew from each other fresh vitality and thenceforward marched on at a rapid pace towards perfection.

Joseph Louis Lagrange.

Bussum, May 1988   Michiel Hazewinkel

·

Table of Contents

Preface

An algebraic-geometric approach to coding theory is developed in this book. Many of the concepts and results of algebraic geometry can be translated nicely into properties of error-correcting codes. Following an account of the historical development of algebraic geometry, codes are constructed. At first, these codes are constructed from conics and rational curves, and subsequently the codes are developed from linear series and the generalized Jacobian variety of a curve. Here, the Riemann-Roch theorem becomes the basic computational tool for estimating the code parameters.

The idea of applying the Riemann-Roch theorem to the analysis of codes was proposed earlier in my article "Codes, associated with divisors" (Problems of Information Transmission, 1977, No. 1, which was written on the occasion of the 25th anniversary of the fundamental Hamming article, introducing error-correcting codes). The present book can be considered as a realization of that idea.

The book does not presume the reader's familiarity with coding theory for algebraic geometry.

I am grateful to Professor M. Hazewinkel for his kind invitation to publish such a book in "Mathematics and Its Applications" series.

Moscow *V.D. Goppa*

CHAPTER 1

Rational Codes

1. Notion of Error-Correcting Code

Let A be an arbitrary finite set (alphabet) consisting of $q = |A|$ letters. By A^n we will denote the set of all sequences (words) of length n in the alphabet A. The Hamming distance $\rho(x,y)$, with x, y belonging to A^n, is the number of positions in which the words x and y are distinguished. It is easy to check that this function satisfies all distance axioms:

1) $\rho(x,y) \geqslant 0$;
2) $\rho(x,y) = 0 => x = y$
3) $\rho(x,z) \leqslant \rho(x,y) + \rho(y,z)$.

The Hamming metric has turned out to be a convenient means of describing noise effects in engineering. If a word x transmitted through a communication channel is received at the output as a word y, with $\rho(x,y) = d$, it is said that d errors have occurred. In order to make it possible to detect and correct errors of a certain multiplicity, usually not all possible q^n words but only some selected "code" words are transmitted. A set of such words is called code U. The key parameters of the code are length n, cardinality $|U|$ and minimum distance of code, $d(U) = \min(x,x')$, $x \neq x'$. The value $R = \dfrac{1}{n} \lg_q |U|$ is called transmission rate and $(1 - R)$ the redundancy of the code. The correcting ability of a code is defined through its minimum distance d and frequent use is made of the relative value d/n. If $d(U) = 2t + 1$ the code corrects t errors. Indeed, in this case all spheres with radius less than or equal to t and centered at code words, do not intersect, hence if an error, whose multiplicity is less than or equal to t occurs, it can always be unambiguously determined what word has been transmitted. For this purpose a code dictionary (that is, all code sequences) is stored at the output and the received word y is compared in turn with all code words. Once $\rho(x,y)$ less than or equal to t occurs for a code word x, the decision is made that word y was transmitted. In other words, the word y is decoded into the code word x for which $\rho(x,y)$ is minimal.

If the minimal admissible transmission rate R is set, that is, redundancy is fixed, the natural requirement is to make optimal use of this redundancy: it is necessary to choose U words in such a way that the paired minimum distance of codes would be the greatest possible one.

In other practical situations the level of noise in the channel is known and it is required to ensure the desirable transmission reliability d/n. In this case, given n and d, it is necessary to find the code of maximum power (the greatest possible dense sphere packing in the Hamming space). Besides, practical applications call for the effective specification of a code dictionary and the construction of fast decoding algorithms.

2. Linear Codes

If $q = p^m$, p is a prime number, the symbols of the alphabet can be regarded as elements of a finite field F_q and the set A^n as a vector space over this field. A linear subspace of A^n is a linear code.

At first glance the condition $q = p^m$, which opens up the possibility of transition to linear codes, appears to be too restrictive. In reality binary codes $(q = 2)$ are of the greatest interest both in terms of practice and theory, so in this case theory always exists the possibility of linearization.

Let U be a k-dimensional linear space. The value k denotes the number of the information symbols of a code and the value $R = n - k$ the number of check symbols. These definitions derive from the fact that any linear code can be reduced to a so-called systematic form, which enables us to obtain a code word in the following manner: an arbitrary selection is made of k first coordinates $(a_1, a_2, ..., a_k)$ and r redundant symbols which are linear combinations of the first k coordinates are cancatenated to them:

$$a_{k-1} = \sum_{j=1}^{k} c_{1j} a_j$$

....

$$a_{k-r} = \sum_{j=1}^{k} c_{rj} a_j.$$

(Here all $a_i, c_{ij} \in F_q$.)

Indeed, the linear code is a null-space of an $r \times n$-matrix H (it is called a parity check matrix of a code).

By a linear transformation of rows and a permutation of the columns this matrix can be reduced to the following canonical form

$$[P, I_r]$$

where P is an $r \times k$-matrix and I_r is a unit $r \times r$ matrix. It is easily verified that $I_k, -P^T$ is also the generating matrix of the code, that is, all its rows make up the basis of the linear space U.

So linear codes actually solve the problem of effective coding: if an arbitrary code dictionary of size q^k requires an exponential memory $q^k \times n$, the linear code dictionary of the same volume needs only a memory of $n \times k$. In fact, it suffices to store the generating matrix of the code.

Let x belong to A^n. The number of nonzero coordinates of vector x in the fixed basis A^n is called the Hamming weight of word x. This weight is a norm on A^n, therefore

$$d(x + y) \leqslant d(x) + d(y).$$

If $x,y \in A^n$, then $\rho(x,y) = d(x-y)$. It follows that for a linear code the minimum distance coincides with the minimum weight of the code's nonzero vectors:

$$d(U) = \min_{x \in U} d(X), \; x \neq 0.$$

This number is sometimes called the weight of a code. So for a linear code the task of calculating the minimum distance is simplified: instead of scanning all possible pairs (x,y) of code words it is sufficient to scan all such words one by one. The following statement is easily verified. If the weight of the code is greater than or equal to d, then any $d-1$ columns of the parity check matrix are linearly independent; the converse is also true.

Since $r + 1$ columns are always dependent, the upper bound holds:

$$d \leqslant r + 1. \tag{1}$$

If U is a linear code (n,k) then the dual space U^* is called an $(n, n-k)$ code which is dual to U. If G and H are the generating and parity check matrices of the code U, then H and G are the respective generating and parity check matrices of code U^*. In the general case attempts to establish any relationships whatsoever between the weights $d(U)$ and $d(U^*)$ have failed. But there exists a simple functional relation between the spectra of U and U^*. The spectrum of a code is the distribution of weights $\{N_0, N_1, ..., N_n\}$, where N_i is the number of code words of weight i. The generating function

$$N(z) = \sum_{i=1}^{n} N_i z^i$$

is called the enumerator of the weights. If $N^*(z)$ is the enumerator of the dual code then

$$N(z) = q^{k-n}(1+(q-1)z)^n N^*((1-z)/(1+(q-1)z)).$$

This relation, called the McWilliams identity, can be proved in the following manner.

We fix j components (coordinates) of a vector. The set of all vectors in which these j coordinates are equal to zero evidently is a subspace. By selecting various positions of these j coordinates one can construct $\binom{n}{j}$ subspaces $S^{(m)}$, $1 \leqslant m \leqslant \binom{n}{j}$.

Let us denote by $S_*^{(m)}$ the complementary subspace to $S^{(m)}$. Each vector of $S_*^{(m)}$ has $(n-j)$ zero coordinates which are complementary to the j fixed coordinates in $S^{(m)}$. For any vector x we put $|x \cap S^{(m)}| = 1$, if x belongs to $S^{(m)}$ and $|x \cap S^{(m)}| = 0$ otherwise. Let $d(x) = i$ and $n - i \geqslant j$. Then

$$\sum_{m=1}^{\binom{n}{j}} |x \cap S^{(m)}| = \binom{n-i}{j}$$

(we select j components from $n - i$ zero coordinates of vector x.)

Summing up all vectors of code U we obtain:

$$\sum_{x \in U} \sum_{m=1}^{\binom{n}{j}} |x \cap S^m| = \sum_i N_i \binom{n-i}{j} = \sum_{m=1}^{\binom{n}{j}} |U \cap S^{(m)}|.$$

Similarly, we have:

$$\sum_{x \in U^*} \sum_{m=1}^{\binom{n}{j}} |x \cap S_*^{(m)}| = \sum_i N_i^* \binom{n-i}{n-j} = \sum_{m=1}^{\binom{n}{j}} |U^* \cap S_*^{(m)}|.$$

Further,

$$dim(U^* \cap S_*^{(m)}) = n - dim(U^* \cap S_*^{(m)})^* =$$

$$= n - dim(U \oplus S^{(m)}) = n - dimU - dimS^{(m)} +$$

$$+ dim(U \cap S^{(m)}).$$

Therefore

$$|U \cap S^{(m)}| = |U^* \cap S_*^{(m)}| \cdot q^{k-j}.$$

So

$$\sum_i N_i \binom{n-i}{j} = q^{k-j} \sum_i N_i^* \binom{n-i}{n-j}.$$

In this equality j is an arbitrary integer from the interval $[0,n]$. We multiply both parts by z^j and sum them over j. As a result, on the left we obtain the sum

$$\sum_j \sum_i N_i \binom{n-i}{j} \tilde{z}^j = \sum_i N_i (1+\tilde{z})^{n-i},$$

and on the right

$$\sum_j q^{k-j} \sum_i N_i^* \binom{n-i}{n-j} \tilde{z}^j = q^k (\frac{\tilde{z}}{q})^n \sum_i N_i^* (1+\frac{q}{\tilde{z}})^{n-i}.$$

Upon dividing both sides by $(1+\tilde{z})^n$ and performing the substitution $z = 1/(1+\tilde{z})$ we arrive at the desired identity.

3. Decoding

Let x be an arbitrary code vector and y the corresponding received vector. The vector y results from x in the presence of noise $y = x + e$. The vector e is usually referred to as the error vector. To obtain the received vector y, the so-called syndrome S is evaluated: $S = Hy$, where H is the parity check matrix of a code. If $S = 0$, it is decided that interference equal zero (by code definition $Hx = 0$), whereas if $S \neq 0$, an error correction procedure is initiated. Clearly $S = H(x+e) = He$ so the Hamming minimum distance decoding taken in the case of linear codes the following form: we know the coset $U + e$ relative to the linear space U; the task is to find a vector e of least weight in the coset. If $d(U) = 2t + 1$, all the vectors with weights below t lie in different cosets and errors of multiplicity below t are corrected unambiguously. But even if $d(e) > t$ this error can be corrected when e is the only vector of least weight in its coset (a coset leader).

Therefore, any noise vector e belongs to A^n results in erroneous Hamming decoding if and only if it is not the leader in its coset, i.e., if there is in $U + e$ another vector e', such that $d(e') \leq d(e)$.

Since the coset contains q^k vectors, the task of finding the vector of least weight in it needs exponential computation time. There seems to be no general algorithm for all the linear codes that would reduce substantially this scanning, yet linear codes can be designed so as to provide a simple decoding procedure.

Incidentally, it suffices for efficient decoding to determine by the syndrome the multiplicity of the error. Moreover, it suffices to identify for any pair of error vectors e and e' which of the relations $S(e) < S(e')$, $S(e) = S(e')$, $S(e) > S(e')$ holds for the corresponding syndromes. Let the syndrome S corresponding to the vector $y = x + e$ be evaluated. To decode, let us alter the components of the vector coordinates one by one, determining the new syndrome for every alteration. If we know how to determine by the syndrome if the multiplicity of the error has decreased, we will be able to uniquely correct all the errors in nq steps.

4. Extension of basic field

It is convenient in many cases to set a q-ary linear code by using a space over the extension F_q^m of the F_q field. Binary codes, for instance, are easy to construct with the help of matrices over the F_{2^m} field. The F_q^m field is a finite extension of the m-th degree of the F_q field. The substitution $x \to \sigma x = x^q$ (the Frobenius automorphism) is defined for every $x \in F_{q^m}$. All these substitutions $\sigma, \sigma^2, \ldots, \sigma^{m-1}$, $\sigma^m = 1$ form the cyclic groups of order m (the Galois field of the group F_{q^m} over F_q). Similarly, we can consider the Frobenius transformation of the $x \in F_q^n{}_m$ (the σ operator applies to all components of the vector).

A vector all components of which belong to the F_q field is called rational over F_q. Every such f vector is determined by the conditions $\sigma x = x$, i.e., stands up to all automorphisms of the Galois group.

Let the parity check matrix H_{q^m} be defined over the F_{q^m} field. Let us identify among the vectors of the code U_{q^m} all vectors rational over F_q. These vectors determine a new code U_q, now over the field F_q (it is known as a subcode over a subfield). The parity check matrix H_q of the U_q code is obtained from the H_{q^m} matrix by the addition to the original rows $x_1, \ldots, x_r$ of new rows

$$\sigma x_1, \ \sigma^2 x_1, \ldots, \sigma^{m-1} x_1, \ \sigma x_2, \ldots, \sigma^{m-1} x_2, \ldots, \sigma x_r, \ldots, \sigma^{m-1} x_r.$$

In the general case, when there are no conjugate vectors among the vectors of the basis $\{x_1, \ldots, x_r\}$, we obtain the bound $r' \leqslant mr$ on the number of the check symbols of the code U_q.

The following rows of the matrix H_q can be taken as a basis rational over F_q:

$$\{Tr(x_1\omega_1), T_r(x_1\omega_2),..., Tr(x_1\omega_m), Tr(x_2\omega_1),...\}.$$

Here $(\omega_1,...,\omega_m)$ is the basis of the field F_{q^m}/F_q,

$$Tr\alpha = \alpha + \sigma\alpha +... + \sigma^{m-1}\alpha,$$

Trx is a vector such that the trace operator is applied to every component.

The main advantage of the above pattern subcode over a subfield is that code weight U_{q^m} usually is easy to determine and yields the estimate $d(U_q) \geqslant d(U_q m)$ for the weight of the code U_q. In most cases this estimate is too rough as weight grows considerably as we pass to the subcode over a subfield.

The q-ary code can be obtained from the q^m-th code also in the following way. Let a vector $x \in U_q m$ have the form $x = (a_1, a_2,..., a_n)$, where all a_i belong to $F_q m$. Each element a_i can be represented as a q-ary vector $\bar{a}_i$ of length m by expanding a_i in the basis $\omega_1,...,\omega_m$. Then a vector $\bar{x}$ from the q-ary code of length $n \cdot m$ corresponds to the x vector. Finally, if a q-ary (n',m)-code V has been set, the vector $\bar{a}_i$ can be extended to a vector b_i with length n', considering the components of $\bar{a}_i$ as information symbols of the code V. The mapping

$$x = (a_1,..., a_2,..., a_n) \rightarrow (b_1, b_2,..., b_n)$$

determines a q-ary (nn', km)-code, whose weight is not less then $d(U_q m) \cdot d(V)$. The $U_q m$ code is usually referred to as external and the V code as internal. This method of combining two codes is known as concatenation. The parity check matrix of the resultant code is the Kronecker product of the parity check matrices used in concatenation.

There are two different types of isomorphism for linear codes. Weak isomorphism consists in that the basic code parameters n, k and d coincide, whereas it is the code spectra that coincide in the case of strong isomorphism. In other words, two linear codes are strongly isomorphic if there is a one-to-one weight preserving between them. In particular, the permutation of the columns of the parity check (generating) matrix which leaves the code invariant is automorphism. The set of such permutations is known as the symmetry group of a code.

5. Hamming code correcting a single error

The code correcting a single error is the simplest nontrivial case of a linear code. In this case the code weight is 3, therefore any two columns of the error-checking matrix must be linearly independent. This means that any two columns cannot differ from one another only by a factor of the field F_q.

We shall represent every column as a vector $(x_0,...,x_{r-1})$ of an affine space. If we omit zero vector in A^r and identify the vectors $X = (x_0,...,x_{r-1})$ and $X' = \alpha X$, $\alpha \neq 0$, then we obtain a projective space P^{r-1}. Thus, in order to obtain the longest code of weight 3, one should incorporate all points of the space P^{r-1} into the error-checking matrix. The corresponding code is called the Hamming code.

Example. $q = 2$, $r = 3$

$$H = \begin{bmatrix} 0 & 0 & 1 & 0 & 1 & 1 & 1 \\ 0 & 1 & 0 & 1 & 0 & 1 & 1 \\ 1 & 0 & 0 & 1 & 1 & 0 & 1 \end{bmatrix}.$$

The decoding is trivial in this case. Let the error vector be of the form $(\overset{0}{1}, \overset{0}{2}, ...\overset{\alpha}{j}...\overset{0}{n})$. On multiplying the matrix H by this vector, we obtain the syndrome $S = \alpha X_j$, where X_j is the j-th column of the matrix H. Therefore, the syndrome is equal to the number of the position where error occurs to within a multiplier showing the magnitude of the error.

For an arbitrary code with the separation $d = 2t+1$ all the spheres of radius t centred at code words do not intersect, and since the volume of such a sphere is exactly $\sum_{i=0}^{t} \binom{n}{i}(q-1)^i$, it follows that the Hamming upper bound holds for every code:

$$|U| \leqslant q^n / \sum_{i=0}^{t} \binom{n}{i}(q-1)^i$$

$$r \geqslant \log_q \sum_{i=0}^{t} \binom{n}{i}(q-1)^i$$

and this converts to equality if all the spheres fill the set A^n throughout. This code is called perfect or densely packed. It is easily checked that the Hamming codes are perfect with the parameters

$$t = 1, \quad n = (q^r - 1)/(q - 1).$$

6. Ovals in the Projective Plane

We now discuss codes with separation $d = 4$. These codes correct one error and detect two. If the separation is $d = 3$, then all errors of multiplicitly greater than 1 cannot be detected and are translated as single error, therefore in this case errors are replicated as the multiple errors are decoded. But if $d = 4$, then using the syndrome one can find out whether one error or two errors occurred, a single error is corrected, and if the error is of multiplicity two, then the corresponding signal appears, but the correction is not performed (in order to correct all errors of multiplicity 2, the code separation must be at least 5).

If $d = 4$, then any 3 columns of the error-checking matrix must be linearly independent, i.e., the corresponding points of the space P^{r-1} must be noncollinear (cannot lie on the same straight line). A maximum amount of points of the projective plane P^2 with such a property is called an oval.

Thus, the notion of Hamming code with separation $d = 4$ (which are often called extended Hamming codes) is equivalent to the well-known geometrical notion of oval.

We first find out how many points can lie on an oval, i.e., which is the maximum length n of the corresponding code.

To this end we note that any straight line intersects an oval either twice (a chord), or a single point (a tangent), or it has no common points with the oval at all (an outside line).

All in all there are $q^2 + q + 1$ points in the plane P^2 and the same quantity of the straight lines. Every line passes through $q + 1$ points and every point is incident to $q + 1$ lines.

We choose an arbitrary point P on the oval U. On joining this point with all remaining $(n - 1)$ points of the oval, we obtain $n - 1$ different straight lines. It follows that $n \leqslant q + 2$ and there are $t = q + 2 - n$ tangents passing through the point P.

In order to derive a more accurate inequality for n, we now choose a point 0 which does not belong to the oval, and denote by $i = i(0)$ the number of chords passing through 0 and by $j = j(0)$ the number of tangents passing through 0.

Then

$$2i + j = n, \text{ hence } j \equiv n \pmod 2$$

$$t = q + 2 - n, \text{ hence } t \equiv n \pmod 2$$

q is even and $t \neq n \pmod 2$, whenever q is odd.

If $t = 0$, i.e., there are no tangents to the oval, then $j = 0$ as well, therefore, simultaneously $n \equiv 0 \pmod 2$ and $n \neq 0 \pmod 2$ for odd q. Consequently, in this case $t \geqslant 1$.

Therefore,

$$n \leqslant q + 1 \text{ if } q \text{ is odd,}$$

$$n \leqslant q + 2 \text{ if } q \text{ is even.}$$

We now discuss the problem of building the ovals (i.e., the extending Hamming codes).

We first consider the oval U for an odd q. In this case, as we have just shown, there exists a tangent at every point of the oval. At the outside circumscribed triangle inscribed in the oval and any circumscribed triangle are in perspective position. To be more precise, let A_1, A_2, and A_3 be three points of the oval (by definition the points do not lie on the same straight line). Let a_1, a_2, and a_3 be the tangents to the oval at the points A_1, A_2, and A_3 respectively. Let A_{12}, A_{13}, and A_{23} denote the intersection points of a_1 and a_2, a_1 and a_3, a_2 and a_3. Those points are the vertices of the circumscribed triangle. Join A_{12} and A_3, A_{13} and A_2, and A_{23} and A_1. It turns out that the three lines intersect at the same point K which is the centre of perspectivity of both triangles (Figure 1).

Let (x_1, x_2, x_3) be homogeneous coordinates in P^2. Without loss of generality we can assume that A_1, A_2, and A_3 coincide with the vertices

$$A_1: (1,0,0); \; A_2: (0,1,0); \; A_3: (0,0,1)$$

since a triple of noncollinear points can always be converted to another one by a projective transformation.

The tangents at those points are of the form

$$a_1: x_2 = k_1 x_3; \; a_2: x_3 = k_2 x_1; \; a_3: x_1 = k_3 x_2,$$

where k_1, k_2, k_3 are three nonzero elements of the field F_q.

Let $B: (c_1, c_2, c_3)$ be any of $(q-2)$ points of the oval different from A_1, A_2, A_3. The straight lines A_1B, A_2B, and A_3B are described by the equations

$$x_2 = \lambda_1 x_3, \; x_3 = \lambda_2 x_1, \; x_1 = \lambda_3 x_2,$$

where

$$\lambda_1 = c_2 c_3^{-1}, \; \lambda_2 = c_3 c_1^{-1}, \; \lambda_3 = c_1 c_2^{-1}.$$

Thus

$$\lambda_1 \lambda_2 \lambda_3 = 1. \tag{1.1}$$

Conversely, if λ_1 is any of $(q-2)$ elements of the field F_q different from 0 and k_1, then the line $x_2 = \lambda_1 x_3$ intersects the oval at A_1 and at some other point, say B, which differs from A_1, A_2, and A_3. Therefore, the coefficients λ_2 and λ_3 in the equations $x_3 = \lambda_2 x_1$ and $x_1 = \lambda_3 x_2$ describing the lines A_2B and A_3B are functions of λ_1 related by condition (1.1) and assuming exactly one nonzero value different from k_2 and k_3 respectively. On multiplying the $q-2$ equations (1.1), we obtain

$$\Pi^3 = k_1 k_2 k_3,$$

where Π is the product of all nonzero elements of the field F_q (the roots of the equation $X^{q-1} - 1 = 0$). Thus

$$k_1 k_2 k_3 = -1. \tag{1.2}$$

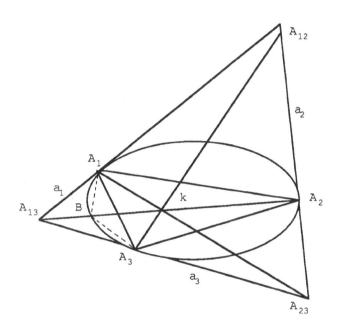

Figure 1

Furthermore, it is easily seen that the lines $A_1 A_{23}$, $A_2 A_{13}$, and $A_3 A_{12}$ are described by the equations

$$x_3 = k_2 k_3 x_2$$

$$x_1 = k_3 k_1 x_3$$

$$x_2 = k_1 k_2 x_1$$

and, therefore, intersect at a single point

$$K: (1, k_1 k_2, -k_2),$$

as we wished to prove.

Note that the condition: three lines described by the equations

$$c_{i1}x_1 + c_{i2}x_2 + c_{i3}x_3 = 0, \quad i = 1, 2, 3$$

intersect at a single point, is equivalent to the conditions: three points lie on a single straight line in view of the duality of the projective plane. Algebraically, this fact is expressed as

$$\|c_{ij}\| = 0, \quad i,j = 1, 2, 3. \tag{1.3}$$

We choose $K = (1,1,1)$. This corresponds to the values $k_1 = k_2 = k_3 = -1$.

Furthermore, we denote by b the tangent to B

$$b: b_1x_1 + b_2x_2 + b_3x_3 = 0. \tag{1.4}$$

The triangles BA_2A_3 and ba_2a_3 are in perspective position, therefore

$$\begin{vmatrix} c_3 - c_2 & c_1 + c_3 & -c_1 - c_2 \\ b_1 - b_3 & b_2 & 0 \\ b_1 - b_2 & 0 & b_3 \end{vmatrix} = 0$$

and it follows that

$$b_2(c_1 + c_2) = b_3(c_1 + c_3).$$

Similarly, considering the triangles BA_3A_1 and BA_1A_2 and the corresponding circumscribed triangles, we obtain the conditions

$$b_3(c_2 + c_3) = b_1(c_2 + c_1), \quad b_1(c_3 + c_1) = b_2(c_3 + c_2).$$

The three equations imply

$$b_1 : b_2 : b_3 = (c_2 + c_3) : (c_3 + c_1) : (c_1 + c_2),$$

and it easily follows that

$$c_2c_3 + c_3c_1 + c_1c_2 = 0.$$

This means that any point of the oval lies on the curve of the second order

(the conic)

$$x_1x_2 + x_1x_3 + x_2x_3 = 0.$$

This proves the Segre theorem which gives a complete description of the oval structure for odd q.

Theorem 1. *If $p \neq 2$, then any oval over the field F_q, $q = p^m$, is a conic (i.e., it can be described by an irreducible equation of the second degree).*

The term 'irreducible' means that the form

$$F(x_1, x_2, x_3) = a_1 x_1^2 + a_2 x_2^2 + a_3 x_3^2 + \tag{1.5}$$

$$+ a_{12}x_1x_2 + a_{13}x_1x_3 + a_{23}x_2x_3.$$

cannot be written as the product of two straight lines. For example, the form $x_1x_3 + x_1x_2 + x_2x_3$ is irreducible since it is linear in x_1 with coefficients of the filed $k(x_2, x_3)$. Let us cross this conic with a pencil of lines passing through the point $A_3 = (0,0,1)$ (which obviously lies on this conic). The equation of the pencil is of the form

$$x_2 = tx_1, \quad t \in F_q. \tag{1.6}$$

On substituting (1.6) into the equation of the curve, we obtain

$$x_1(tx_1 + (t + 1)x_3) = 0.$$

If $x_1 = 0$, the $x_2 = 0$ as well, and the initial point A_3 comes up. The second point of intersection of the line and the conic is defined by the parametric equation

$$x_2 : x_1 = t$$

$$x_3 : x_1 = -t / (t + 1) \quad \text{if } t \neq -1$$

$$x_1 = x_2 = 0, \quad x_3 = 1 \text{if } t = -1.$$

As t runs the field F_q, we obtain q different points of the conic. Finally, the line $x = 0$ intersects the curve at the point $A_2 = (0,1,0)$. Therefore, there is a one-to-one correspondence between lines of the pencil and $q + 1$ points of the conic, and the coordinates of the conic are expressed as a rational function of t. The parameter t itself can also be represented as a rational function of the coordinates of a point, i.e., $t = x_2/x_1$.

The same is true for any irreducible conic described by the general equation (1.5). This can easily be shown by crossing the conic with the pencil of lines passing through a point of this conic.

We consider the set of functions corresponding to the coordinates of a point of the curve

$$x_1 = 1,$$

$$x_2 = t, \tag{1.7}$$

$$x_3 = -t/(t + 1).$$

On carrying out the projective transformation of the plane

$$x_1' = a_{11}x_1 + a_{12}x_2 + a_{13}x_3$$

$$x_2' = a_{21}x_1 + a_{22}x_2 + a_{23}x_3 \tag{1.8}$$

$$x_3' = a_{31}x_1 + a_{32}x_2 + a_{33}x_3,$$

with a nonsingular matrix $\|a_{ij}\|$ we obtain a new irreducible conic which is projectively equivalent to the initial one, the parametrization being changed as

$$x_1' = f_1(t)$$

$$x_2' = f_2(t), \tag{1.9}$$

$$x_3' = f_3(t).$$

Since in a projective space the points are considered to within the multiplication by a constant, it follows that all functions in (1.7) can be multiplied by $(t + 1)$. It is readily seen that $\{t + 1, t(t + 1), -t\}$ make up a basis in the space of all polynomials in t of the second degree. Therefore, the function $f_1(t)$, $f_2(t)$, $f_3(t)$ make up a basis as well. It follows that the irreducible conics are projectively equivalent and consist of $q + 1$ points.

As a standard parametrization one can take

$$x_1 = 1,$$

$$x_2 = t \quad (x_1 x_3 = x_2^2), \tag{1.10}$$

$$x_3 = t^2.$$

In the context of coding theory, the Segre theorem is to show how algebraic curves come up in connection with codes. Moreover, the theorem is an example of 'classification' theorems in coding theory since it provides the description of all extended Hamming codes with 3 control symbols for an odd q.

In the case of even q, every line intersects the irreducible conic only at 2 points hence every 3 points are noncollinear. But in such a case all tangents to the conic intersect at the same point called a centre.

It suffices to check this property for the conic $x_1 x_3 = x_2^2$.

The equation of the tangent at the point (x_1^0, x_2^0, x_3^0) is

$$x_3(x_1 - x_1^0) + x_1(x_3 - x_3^0) = 0.$$

Thus all the tangents pass through the point $(0,1,0)$, which is the centre.

On adding this point to other points of the conic we obtain an oval consisting of $q + 2$ points.

As an example we construct an extended Hamming code with $q = 4$, i.e., an oval of length 6 and the corresponding (6,3,4) code. (Here we use the abbreviated notation of parameters (n,k,d).)

The field F_4 is described by the irreducible polynomial $x^2 + x + 1$ and consists of 4 elements 0, 1, α, β. Here $\alpha + \beta = 1$, $\alpha + 1 = \alpha^2 = \beta$, $\beta + 1 = \beta^2 = \alpha$.

$$
\begin{array}{c|cccccc}
x_1 = 1 & 1 & 1 & 1 & 1 & 0 & 0 \\
x_2 = t & 0 & 1 & \alpha & \beta & 0 & 1 \\
x_3 = t^2 & 0 & 1 & \beta & \alpha & 1 & 0 \\
\hline
 & 0 & 1 & \alpha & \beta & &
\end{array}
$$

The last column is the centre of the oval and the next to the last column corresponds to the infinite point of the conic ($x_1 = 0$).

The decoding is carried out in the following manner: the syndrome is compared with all the columns of the error-checking matrix. If the syndrome turns out to be proportional to the j-th column for a j, this means that an error has occurred in the j-th position and the proportionality factor is equal to the magnitude of the error. But if there is no such a j, this is the evidence that an error of multiplicitly 2 has occurred. In this case no correction is made and a signal is generated showing that an error has occurred which cannot be corrected and the transmission should be repeated.

7. Codes on a Twisted Cubic Curve

We now construct codes correcting two errors ($d = 5$). If $r = 3$, there are no such codes. Indeed, if $d = 5$, then any 4 columns of the error-checking matrix must be linearly independent, which is impossible in the projective plane.

Thus we consider the case $r = 4$. A set of n points of the space P^3 such that any 4 points are not coplanar (do not lie in the same plane) is called the n-arc.

Consequently, a linear code with separation 5 corresponds to an arc in P^3. We choose an arbitrary point Q_1 of the arc and project arc on a plane which does not pass through Q_1. To this end we joint the point Q_1 with a point P of the arc. Then the line PQ_1 intersects the plane at a point P' which is the projection of P.

The resulting set of q points $\{P'\}$ is such that no 3 of those points lie on the same line. Similarly as in the proof of the Segre theorem on ovals, one can show that any set like that consists of points of an irreducible conic. In other words, one more point should be added to this set to obtain a conic. The set of lines connecting the point Q with points of this conic make up a cone (a conic surface). Choosing another point Q_2 of the arc, we obtain another conic surface. Without loss of generality, one may assume that $Q_1 = (0,0,0,1)$, $Q_2 = (1,0,0,0)$ and the projection plane in the first case is $x_4 = 0$, and in the second case $x_1 = 0$. Then the cones under consideration are described by the equations

$$x_1 x_3 = x_2^2, \tag{1.11}$$

$$x_2 x_4 = x_3^2.$$

The intersection of these quadrics is a rational curve having the parametric representation

$$x_1 : x_2 : x_3 : x_4 = t^3 : t^2 : t : 1. \tag{1.12}$$

These quadrics both have a straight line (generator) $x_2 = 0 = x_3$. Therefore, the intersection which is a curve of degree 4 fall into a straight line and an irreducible curve of degree 3, which is called the twisted cubic. More generally the twisted cubic is defined as a curve with the parametric representation

$$x_1 : x_2 : x_3 : x_4 = f_1(t) : f_2(t) : f_3(t) : f_4(t),$$

where f_i are linearly independent polynomials of degree not greater than 3.

Using a projective transformation, this curve can be reduced to the canonical form (1.12), hence all the twisted cubics are projectively equivalent.

Since every quadric contains the initial arc, it follows that their intersection contains the arc. However, the twisted cubic itself is a $(q + 1)$-arc since every its plane has no more than 3 points of the cubic. Indeed, if a point of this curve corresponding to $t = t_1$ lies in the plane

$$u_1 x_1 + u_2 x_2 + u_3 x_3 + u_4 x_4 = 0,$$

then it means that by (1.12)

$$u_1 t_1^3 + u_2 t_1^2 + u_3 t_1 + u_4 = 0. \tag{1.13}$$

Since an equation of degree 3 cannot have more than 3 roots, every plane intersects the curve at no more than 3 different points. In this case the curve is said to have degree 3. If t_1, t_2, t_3 are parameters of 3 points of the curve, then the plane passing through these points is described by the equation

$$\frac{u_1}{1} = \frac{u_2}{-(t_1+t_2+t_3)} = \frac{u_3}{t_1 t_2 + t_1 t_3 + t_2 t_3} = \tag{1.14}$$

$$= \frac{u_4}{-(t_1 t_2 t_3)}.$$

This equation will be used for decoding. Here we formulate the Segre result.

Theorem 2. *For odd q, any $(q + 1)$-arc in the space P^3 is a twisted cubic. Such an arc can have no more than $q + 1$ points.*

Thus we obtain a complete description of linear codes correcting 2 errors for odd q. For even q the twisted cubic naturally leads to the same codes since this cubic is a $(q + 1)$-arc.

For example, let $q = 2^4 = 16$. In the field F_{16} operations will be performed modulo the polynomial $X^4 + X + 1$. If α is a root of this polynomial, then the table of calculation is as follows

	α^3	α^2	α	1			α^3	α^2	α	1
$\alpha^0 =$				1	$\alpha^8 =$			α^2		1
$\alpha^1 =$			α		$\alpha^9 =$	$\alpha^3 +$			$+\alpha$	
$\alpha^2 =$		α^2			$\alpha^{10} =$		$\alpha^2 +$		$\alpha +$	1
$\alpha^3 =$	α^3				$\alpha^{11} =$	$\alpha^3 +$	$\alpha^2 +$		α	
$\alpha^4 =$			$\alpha +$	1	$\alpha^{12} =$	$\alpha^3 +$	$\alpha^2 +$		$\alpha +$	1
$\alpha^5 =$		$\alpha^2 +$	α		$\alpha^{13} =$	$\alpha^3 +$	$\alpha^2 +$			1
$\alpha^6 =$	$\alpha^3 +$	α^2			$\alpha^{14} =$	α^3			$+$	1
$\alpha^7 =$	$\alpha^3 +$		$\alpha +$	1	$\alpha^{15} =$					1

The twisted cubic is described by the parametric equations

$$x_1 = 1, \ x_2 = t, \ x_3 = t^2, \ x_4 = t^3.$$

We construct the binary code based on this cubic by using the subcode over the subfield according to the method described in Section 4.

Since the functions t and t^2 are conjugate, it follows that the error-checking matrix of the code is defined by the basic functions $\{1, t, t^3\}$:

$$
\begin{array}{c|cccccccc}
1 & 1 & 1 & 1 & 1 & 1 & 1 & 1 & 1 \\
t & 0 & 1 & \alpha & \alpha^2 & \alpha^3 & \alpha^4 & \alpha^5 & \alpha^6 & \alpha^7 \\
t^3 & 0 & 1 & \alpha^3 & \alpha^6 & \alpha^9 & \alpha^8 & 1 & \alpha^3 & \alpha^6 \\
\hline
 & 0 & 1 & \alpha & \alpha^2 & \alpha^3 & \alpha^4 & \alpha^5 & \alpha^6 & \alpha^7
\end{array}
$$

$$
\begin{array}{cccccccc}
1 & 1 & 1 & 1 & 1 & 1 & 1 & 0 \\
\alpha^8 & \alpha^9 & \alpha^{10} & \alpha^{11} & \alpha^{12} & \alpha^{13} & \alpha^{14} & 0 \\
\alpha^9 & \alpha^{12} & 1 & \alpha^3 & \alpha^6 & \alpha^9 & \alpha^{13} & 1 \\
\hline
\alpha^8 & \alpha^9 & \alpha^{10} & \alpha^{11} & \alpha^{12} & \alpha^{13} & \alpha^{14} & \infty
\end{array}
$$

Decomposing every element of the field F_{2^4} over the field F_2, we obtain the binary matrix

1	1	1	1	1	1	1	1	1	1	1	1	1	1	1	1	0
0	0	0	0	1	0	0	1	1	0	1	0	1	1	1	1	0
0	0	0	1	0	0	1	1	0	1	0	1	1	1	1	0	0
0	0	1	0	0	1	1	0	1	0	1	1	1	1	0	0	0
0	1	0	0	0	1	0	0	1	1	0	1	0	1	1	1	0
0	0	1	1	1	0	0	1	1	1	1	0	1	1	1	1	0
0	0	0	1	0	1	0	0	1	0	1	0	0	1	0	1	0
0	0	0	0	1	0	0	0	0	1	1	0	0	0	1	0	0
0	1	0	0	0	1	1	0	0	0	1	1	0	0	0	1	1

Thus we obtain binary codes with parameters

$$n = 2^m + 1, \quad r = 2m + 1, \quad d = 6.$$

The Hamming bound implies that for $n = 2^m$ any code correcting two errors must have at least $2m$ control symbols. Therefore, the binary codes resulting from the twisted cubic are optimal.

We now describe the decoding of such codes. The main idea is that exactly one chord of the curve passes through any point S of the space P^3. Indeed, if $\lambda_1 P_1 + \mu_1 Q_1 = P = \lambda_2 P_2 + \mu_2 Q_2$, then the four points P_1, Q_1, P_2 and Q_2 of the curve are linearly dependent, which contracts to the fact that the curve is an arc. Therefore, it suffices for decoding to find this unique chord passing through the point $P = S$ corresponding to the syndrome.

Let the points P_1 and Q_1 be associated with parameters θ_1, θ_2. The chord (θ_1, θ_2) lies in the plane $(\theta_1, \theta_2, \lambda)$ for all values of λ, i.e., the chord is the axis of the pencil of planes

$$x_1 - (\theta_1 + \theta_2)x_2 + \theta_1\theta_2 x_3 + \lambda[x_2 - (\theta_1 + \theta_2)x_3 +$$

$$\theta_1\theta_2 x_4] = 0,$$

in accordance with equation (1.14). In other words, we substitute θ_3 in the equation of the plane passing through 3 points by a formal parameter λ so that (1.14) is satisfied for any values of λ and therefore

$$x_1 - (\theta_1 + \theta_2)x_2 + \theta_1\theta_2 x_3 = 0 = x_2 - (\theta_1 + \theta_2)x_3 + \theta_1\theta_2 x_4$$

is the equation describing the chord which connects the points θ_1 and θ_2. Substituting the point $S = (s_1, s_2, s_3, s_4)$ corresponding to the syndrome for (x_1, x_2, x_3, x_4) we obtain the set of equations

$$s_1 + s_2 p + s_3 q = 0, \tag{1.16}$$

$$s_2 + s_3 p + s_4 q = 0,$$

where $p = -(\theta_1 + \theta_2)$, $q = \theta_1\theta_2$ are the roots of the polynomials

$$f = \theta^2 + p\theta + q,$$

which is called the error locator. Solving equation (1.16), we find the coefficients p q of this polynomial and its roots pointing at the positions of errors. Now it is not difficult to determine the magnitude of errors (the values of distortion) representing the syndrome as the linear combination of the corresponding columns of the error-checking matrix.

8. Normal Rational Codes

The notions of oval and twisted cubic are generalized by the normal rational curve in P^{r-1} with the parametric representation

$$x_0 : x_1 : \ldots : x_{r-1} = f_0(\theta) : f_1(\theta) : \ldots : 1,$$

where f_i are linearly independent polynomials of degree not greater than $r - 1$. By a projective transformation of the coordinates this curve can be reduced to the canonical form

$$x_0 : x_1 : \ldots : x_{r-1} = \theta^{r-1} : \theta^{r-2} : \ldots 1. \tag{1.18}$$

The points of this curve make an arc in the space P^{r-1}, hence any r points do not lie in the same hyperplane. Indeed, all the points lying in a plane satisfy the equation of degree $(r-1)$

$$u_0 \theta^{r-1} + u_1 \theta^{r-2} + \ldots + u_{r-1} = 0, \tag{1.19}$$

where $(u_0,...,u_{r-1})$ is the coordinate vector of the hyperplane and θ is the parameter of the point P of the curve.

Consequently, any hyperplane intersects the curve at $r-1$ points (the degree of the curve is $r-1$). The form 'normal' means that the curve is not a projection of a curve of the same degree from a larger space. The term 'rational' means that all the points of the curve are in a one-to-one correspondence with the points of the projective line P^1 and this correspondence is defined by rational functions.

The corresponding code, whose error-checking matrix is made up of the points of the curve, is of length $n = q + 1$, has r control symbols, and code separation $d = r + 1$, since any r columns of the matrix are independent. Consequently, this code is optimal, it corrects the maximum number of errors for fixed n, r, and q. As far as the maximum length of the code with parameters $d = r + 1$ (the number of points of the corresponding arc in P^{r-1}) is concerned, in contrast to the specific cases of oval and twisted cubic, we do not know the exact value of n. An upper bound of n can be derived using arguments similar to those applying to the number of the points of an oval: we fix $r - 2$ points of the number of the arc and draw a pencil of hyperplanes through these points. All these hyperplanes make up a manifold which is birationally equivalent to P^1, i.e., the pencil is defined by a single parameter which can be chosen arbitrarily (sometimes ∞^1 stands for free choice of the parameter).

Thus there exists $q + 1$ hyperplanes of the pencil with coefficients in F_q. Some of these hyperplanes can be obtained by joining the initial $r - 2$ points with the remaining points of the arc (we obtain different hyperplanes since every hyperplane contains no more than $r - 1$ points). Hence we obtain the estimate

$$n \leqslant q + r - 1.$$

The dual code to the normal rational code consists of the vectors

$$(f(t_1), f(t_2),...,f(t_n)),$$

where $f(t)$ is a polynomial of degree no greater $r - 1$ in accordance with parametrization (1.17), and t_i runs all the points of P^1. Since this polynomial cannot have more than $r - 1$ roots, it follows that the number of roots among the components of the code vector does not exceed $r - 1$. Therefore, the weight $d*$ of the dual code is no less than $n - r + 1$. Thus the parameter of the dual code also satisfy the condition

$$d* = r* + 1,$$

where $r* = n - r$ is the number of control symbols of the dual code.

The binary codes results from the normal rational code as we pass to the subcode over a subfield. In this code $q = 2^m$, $r' \leq mt + 1$, $d \geq 2t + 2$, $n = 2^m + 1$. The error-checking matrix of such a code consists of the points of a rational curve defined by the parametrization

$$x_0 : x_1 : ...x_r' = f_0 : f_1 : ...f_0^2 : f_1^2 : ... \qquad (1.20)$$
$$f_0^{2^i} : f_1^{2^i} : ...$$

This curve is no longer normal in the corresponding space $P^{r'-1}$. However, all the properties of this curve are completely defined by the properties of the initial normal curve.

We now discuss the decoding problem. Let $r = 2t$. Then the code corrects t errors. Given an arbitrary hyperplane in the space $P^{r'-1}$ defined by the components of the vector $(u_0, u_1,...,u_{2t-1})$, the points of intersection of this plane with the curve are given by the roots of the equations

$$u_0 \theta^{2t-1} + u_1 \theta^{2t-2} + ... + u_{2t-1} = 0.$$

Therefore the equation of the hyperplane passing through the points with parameters $\theta_1, \theta_2,...,\theta_{2t-1}$ is

$$\frac{u_0}{1} = \frac{u_1}{\sigma_1} = \frac{u_2}{\sigma_2} = ... = \frac{u_{2t-1}}{\sigma_{2t-1}} \qquad (1.21)$$

where σ_i are elementary symmetric functions

$$\sigma_1 = -(\theta_1 + \theta_2 + ... + \theta_{2t-1})$$
$$\sigma_2 = \theta_1 \theta_2 + \theta_1 \theta_3 + ...$$

$$............ \qquad (1.22)$$

$$\sigma_{2t-1} = -\theta_1\theta_2...\theta_{2t-1}$$

We replace θ_{2t-1} in these equations by a formal parameter λ. The subspace spanned by the points $[\theta_1, \theta_2,...,\theta_{2t-2}]$ lies in the hyperplane $[\theta_1,...,\theta_{2t-2}, \lambda]$ for all values of λ and is the axis of pencil of hyperplanes

$$f_0 + \lambda f_1 = 0, \tag{1.23}$$

where

$$f_0 = x_0 + \sigma_1^{(1)}x_1 + \sigma_2^{(1)}x_2 + ... + \sigma_{2t-2}^{(1)}x_{2t-2} \tag{1.24}$$

$$f_1 = x_1 + \sigma_1^{(1)}x_2 + \sigma_2^{(1)}x_3 + ... + \sigma_{2t-2}^{(1)}x_{2t-1}$$

and $\sigma_i^{(1)}$, $i = 1,...,2t-2$ are elementary symmetric functions of $\theta_1,...,\theta_{2t-2}$. Indeed, substituting (1.24) into (1.23), we obtain (1.22). It follows that $f_0 = 0 = f_1$ is the equation of the subspace $[\theta_1,...,\theta_{2t-2}]$. Applying the same procedure to functions f_0 and f_1 we find that equation of the space $[\theta_1,...,\theta_{2t-3}]$ is of the same form as in (1.24).

On separatedly applying this technique, we finally obtain the equation of the subspace $[\theta_1,...,\theta_t]$:

$$x_0 + \sigma_1^{(t)}x_1 + ... + \sigma_t^{(t)}x_t = 0$$

$$x_1 + \sigma_1^{(t)}x_2 + ... + \sigma_t^{(t)}x_{t+1} = 0 \tag{1.25}$$

$$............$$

$$x_{t-1} + \sigma_1^{(t)}x_t + ... + \sigma_t^{(t)}x_{2t-1} = 0,$$

where $\sigma_i^{(t)}$, $i = 1,...,t$ are elementary symmetric functions of $\theta_1...\theta_t$. These functions are uniquely defined by system (1.25) if $(x_0,...,x_{2t-1})$ is replaced by the syndrome vector $(s_0,...,s_{2t-1})$. Thus (1.25) is the equation of the unique subspace of dimension no greater than $t-1$ passing through $\theta_1,...,\theta_t$ and the given point S, which is the syndrome. If the rank of matrix (1.25) proves to be less than t, this means that less than t errors have occurred, i.e., not all $\theta_1,...,\theta_t$ are different.

Consequently, the decoding reduces to the solution of a linear system, which yields the coefficients of the error locator polynomial. In the next chapter we shall develop another approach to the decoding of such codes, which is based on the Diophantine approximations. This approach leads to a simpler method of constructing the error locator. The method takes advantage of the fact that the matrix of (1.25) has a specific symmetric structure, namely every row is the shifted previous one. These matrices are called Toeplitz (Hankel).

9. Rational Functions

Consider a rational function of one variable over an arbitrary field k:

$$f(z) = \psi(z) / \phi(z),$$

where ψ and ϕ are polynomials of degrees l and s. Let $\psi(z) = \Pi(z - \beta_i)^{l_i}$, $\Sigma l_i = l$, $\phi(z) = \Pi(z - \alpha_j)^{s_j}$, and $\Sigma s_j = s$. The values β_i and α_j lie in an algebraic extension of the field k and called zeros and poles of the function f of multiplicitly l_i and s_j respectively. All the rational functions make up the field k. The valuation of a field is any real-valued function w possessing the following properties:

(1)$w(0) = 0$, $w(a) \geqslant 0$;
(2)$w(a + b) \leqslant w(a) + w(b)$
(3)$w(ab) = w(a) \cdot w(b)$
for any $a, b \in k$.

This definition readily implies the following properties of $w(a)$:
$w(-a) = w(a)$;
$w(a) - w(b) \leqslant w(a - b) \leqslant w(a) + w(b)$;
$|w(a) - w(b)| \leqslant w(a - b)$;
$w(\pm 1) = 1$.

Every field has at least one valuation, namely the trivial one

$$w_0(a) = \begin{cases} 0 & \text{if } a = 0 \\ 1 & \text{if } a \neq 0 \end{cases}$$

For the trivial valuation the triangle inequality takes a stronger form:

$$(2') \quad w_0(a + b) \leqslant \text{Max}\,(w_0(a),\, w_0(b))$$

(the ultra-metric inequality).

Any valuation satisfying inequality (2') is said to be non-archimedean, other-wise the valuation is archimedean.

Thus the valuation $|a|$ of the rational Q is archimedean. In addition to $|a|$ and $w_0(a)$, the field Q has infinitely many p-adic valuations $|a|_p$, which are defined in the following manner

(i) $|0|_p = 0$;

(ii) if $a \neq 0$, then a can be represented uniquely as $a = p^n \dfrac{r}{s}$, where r and s are

not divided by p and n is an integer. Set

$$|a|_p = p^{-n}$$

It is easily seen that $|a_p|$ is a non-archimedean. If q is prime, then

$$|q|_p = \begin{cases} 1/p & \text{if } q = p \\ 1 & \text{if } q \neq p. \end{cases}$$

It follows that the p-adic valuations corresponding to different prime p are different. Thus Q has infinitely many various valuations $|a|$, $w_0(a)$, and $|a|_p$ for different p. These valuations are interrelated by the fundamental identity

$$|a| \cdot \prod_p |a|_p = w_0(a) \text{ for all } a \in k,$$

which is equivalent to the fundamental theorem of arithmetic on the unique fac-torization of an integer into primes.

In the set of polynomials $k[z]$, the unique factorization theorem is also valid, an irreducible polynomials p being the analog of a prime number. The field of rational functions $k(z)$ is the field of quotients of the ring $k[z]$. For any polyno-mial r we denote by $\deg(r)$ the degree of r and by $\mathrm{ord}_p(r)$ the order of r with respect to p, i.e., the largest integer m such that p^m divides r.

Every element of the field $k(z)$ can be uniquely represented as $a = r/s$, where r and s are relatively prime polynomials and s is a nonzero polynomial with unit leading coefficient. Denote

$$\|a\| = \begin{cases} 0 & \text{if } a = 0 \\ \theta^{\deg(s) - \deg(r)} & \text{if } a \neq 0 \end{cases}$$

$$\|a\|_p = \begin{cases} 0 & \text{if } a = 0 \\ \theta^{\text{ord}}{}_p{}^{(r) - \text{ord}}{}_p{}^{(s)} \text{if } a \neq 0 \end{cases}$$

It is easily seen that the two functions are non-archimedean valuations of the field $k(z)$ and $w_0(a)$, $\|a\|$, $\|a\|_p$ for all irreducible polynomials p make up an infinite system of different valuations interrelated by the fundamental identity

$$\|a\| \cdot \prod_p \|a\|_p^{\deg(p)} = w_0(a).$$

Every valuation of the field defines a metric $\delta(a,b) = w(a-b)$. The sequence $a_1, a_2,...$ is called fundamental if $\lim w(a_i - a_j) = 0$ as $i, j \to \infty$. This sequence may have no limit, but using the well-known Cauchy method one can embed any valuation field into a complete field in which every fundamental sequence has the limit: the element of such a field is the class of 'close' fundamental sequences.

The valuations $w_0(a)$, $|a|$, and $|a|_p$ of the field Q lead to different completions Q, R, and Q_p. Every element of the latter field can be represented as the series

$$x = \sum_{n=-s}^{\infty} a_n p^n, \quad 0 \leqslant a_n < p,$$

where a_n are integers.

If $s = 0$, then x is called the p-adic integer. The infinite sum has just a formal meaning, i.e., x is defined by the sequence

$$\{a_0, a_0 + a_1 p, a_0 + a_1 p + a_2 p^2, ...\}$$

of partial sums S_j. The difference $S_i - S_j$ is divided by a large power of p for sufficiently large i and j, which means that these sums become arbitrarily close in the sense of the p-adic norm $|S_i - S_j|_p$.

The fractional p-adic number is defined as a ratio of two integer p-adic numbers. The number can uniquely be represented as a product of p^m by an integer p-adic number such that $a_0 \neq 0$ (here m may be negative).

The completion of the set $k(z)$ with respect to the norm $\|a\|_p$ consists of formal power series

$$S(z) = \sum_{i=-s}^{\infty} a_i(z)p^i, \quad \deg(a_i(z)) < \deg(p).$$

If the initial constant field k is algebraically closed, then all irreducible polynomials are of the form $(z - \alpha)$, therefore $S(z)$ is the Laurent series in a neighbourhood of α:

$$S(z) = \sum_{i=-s}^{\infty} a_i(z - \alpha)^i.$$

Every such a series can be represented uniquely as a product of $(z - \alpha)^m$ by a power series with nonzero lowest term a_0. The latter condition means that the power series is invertible:

$$\frac{1}{a_0 + a_1 z + \ldots} = \frac{a_0^{-1}}{1 - R(z)} = a_0^{-1}(1 + R(z) + R^2(z) + \ldots)$$

and the expression on the right makes sense since $R(z)$ has no absolute term, hence there cannot be infinitely many summands with the same degree of z.

The embedding of an arbitrary rational function into the field of series is as follows . It suffices to consider only the case when $f(z)$ has no pole at $z = \alpha$ (otherwise it could be multiple by a power of $(z - \alpha)$). We evaluate $f(\alpha)$, then $f(z) - f(\alpha)$ is divided by $z - \alpha$, compute the value of $[f(z) - f(\alpha)]/(z - \alpha)$, and so on.

The completion of field $k(z)$ with respect to the norm $\|a\|$ consists of the series

$$x = \sum_{i=-s}^{\infty} a_i(1/z)^i.$$

Thus every point (x, y) of the projective line P^1 generates a valuation, the point $(1, 0)$ being associated with valuation $\|a\|$ and the point $(\alpha, 1)$ with the valuation $\|a\|_{z-\alpha}$.

In order to obtain the expansion of $f(z)$ at the point ∞ one needs to substitute $z = 1/t$ and to expand the resulting function in a neighbourhood of the point $t = 0$. On expanding the function f into series at every point of P^1, we obtain a set of the Laurent series $\{S_p(z)\}$. This set is called the distribution or adele of the function. Since the function has a finite number of poles, there are among the series $S_p(z)$ only a finite number of series with negative powers.

The coefficient a_{-1} in the expansion $S_p(z)$ is called the residue of the function at the point P and denoted by $\text{Res}_P(f)$. Thus every function generates the vector of residues

$$f \rightarrow (\text{Res}_{P_1}(f),\ \text{Res}_{P_2}(f),...)$$

The operation Res is linear, hence

$$\text{Res}_P(af + bg) = a\ \text{Res}_P(f) + b\text{Res}_P(g).$$

We associate with every point P of the projective line an integer n_p and call the formal finite sum $D = \sum n_p P$ the divisor. In particular, every series $S_P(z)$ generates a divisor $n_p P$, where n_p is the smallest number such that $a_{n_p} \neq 0$. This number is called the order of a series (function) and denoted by $\nu_p(f) = \text{ord}_p(f)$. Thus the adele of a function generates the divisor of the function

$$(f) = \sum \text{ord}_p(f)P$$

The positive coefficient in this sum is equal to the multiplicity of a zero and the negative one is the multiplicity of a pole. Since the number of zeros is the same as the number of poles (with taking into account the infinite point), it follows that $\sum \text{ord}_p(f) = 0$. The value $\sum n_p$ the degree of the divisor denoted by $\deg(D)$.

The differential of f is the expression

$$\omega = df = f'dz,$$

where f' is the formal derivative of the function. On differentiating the series $S_p(z)$, we obtain the adele of the differential, the vector of residues, and the divisor. For any differential the sum of residue is 0:

$$\sum_p \text{Res}_p(\omega) = 0.$$

Since the operation Res is linear, it suffices to prove this formula for $f = z^n$ and $f = 1/(z - \alpha)^n$: any function can be represented as a linear combination of such functions breaked into partial fractions. In the first case, the only pole is the infinite point. In order to find the residue at this point, we have to substitute $z = 1/t$ and find the residue at the point $t = 0$. We obtain $\omega = -n \, dt / t^{n+1}$ hence $\mathrm{Res}_0(\omega) = 0$. In the second case $\omega = dz/(z-\alpha)$ has the poles α and ∞ with the residues 1 and -1 respectively.

While for any function we have $\deg((f)) = 0$, for differentials the equality $\deg(\omega) = -2$ holds true. Indeed, since

$$\mathrm{div}\,(\omega) = \mathrm{div}\,(f) + \mathrm{div}\,(dz),$$

it follows that

$$\deg\,(\mathrm{div}\,(\omega)) = \deg\,(\mathrm{div}\,(dz)).$$

The adele of function z consists of the series $\{\alpha + (z-\alpha)\}$ at every finite point and the series $1/t$ at the point ∞. On differential $\{1\}$ and $-1/t^2$, hence dz has a unique pole of multiplicity 2 at the infinite point.

For any divisor $D = \sum n_p P$, the set of functions

$$L(D) = \{f \,|\, (f) \geqslant -D\} = \{f \,|\, v_p(f) \geqslant -n_p \text{ for all } P\}$$

is called the space associated with the divisor D. For example, if $D = \sum P_{\alpha_i}$, then $L(D)$ consists of the functions

$$f = \sum \frac{a_i}{z - \alpha_i}.$$

If we require that these functions have zeros at the points, i.e., those functions pass through the divisor $G = \sum P_{\beta_i}$, then we obtain the space $L(D-G)$. Obviously, $L(D)$ is a linear space for any divisor D and using the decomposition into partial fractions one easily obtain the dimension of this space

$$l(D) = \deg(D) + 1.$$

Similarly, we can define the space of differentials

$$\Omega(D) = \{\omega \,|\, \mathrm{div}\,(\omega) \geqslant D\} = \{\omega \,|\, \mathrm{ord}_p(\omega) \geqslant n_p\}.$$

For the divisors D and G in the previous example, $\Omega(D-G)$ consists of the differentials

$$\omega = \sum \frac{c_j}{z - \beta_j} dz$$

such that α_j are zeros.

10. Differential Representation of Codes

Let A^n be an affine space of dimension n over the field F_q. Denote the divisor composed of all the point of the projective straight line P^1 over the field F_q: $D = \sum P_i$. We introduce the mapping

$$\Omega(-D) \xrightarrow{\phi} A^n$$

$$\phi: \omega \to (\mathrm{Res}_{p_1}(\omega), \mathrm{Res}_{p_2}(\omega), ..., \mathrm{Res}_{p_n}(\omega)).$$

All the differentials of $\Omega(-D)$ have a pole of multiplicitly no greater than 1, hence $\Omega(-D)$ is generated by differentials of the form $\dfrac{a_i}{z - \alpha_i} dz$ and $a_i = \mathrm{Res}_{p_i}(\omega)$, while $\mathrm{Res}_\infty(\omega) = -\sum a_i$ at infinity. The kernel of the mapping ϕ consists of the differentials which have no poles, i.e., it is zero. Therefore, $\Omega(-D)$ is isomorphic to the subspace of A^n consisting of those vectors whose coordinates added together make zero. Obviously, the dimension of this space is

$$\delta(-D) = n - 1.$$

Thus every vector $x = (a_1, ..., a_n)$, $x \in A^n$, $\sum a_i = 0$ can be identified with a differential of the space $\Omega(-D)$, the Hamming weight of x being equal to the number of poles of the differential. The code can be defined through this correspondence as a set of differentials of $\Omega(-D)$.

Specifically, a linear code U is defined as a subspace of $\Omega(-D)$. If we want to make the code separation greater, all the differentials of $\Omega(-D)$ have to have a large number of poles (zeros). We, therefore, make all the differentials of U

pass through a divisor $G = \sum m_Q Q$ such that the points Q belong to an extension of the field F_q. In other words, we choose a polynomial

$$g(z) = \Pi(z - \beta_j)^{m_j}, \quad \beta_j \notin F_q, \quad \sum m_j = m.$$

The coefficients of this polynomial (called the generating polynomial of the code) belong to the field F_q, hence every root β_j enters $g(z)$ together with its conjugate σ_{β_j}. The polynomial $g(z)$ is associated with the divisor $G = \sum m_j Q_j$. We define the code U as the subspace

$$U = \Omega(G - D).$$

It is clear that the dimension of this space (the logarithmic cardinality of the code) is equal to

$$k = n - 1 - \deg(G), \quad \deg(G) = \deg g(z),$$

the number of control symbols (co-dimension) is equal to

$$r = \deg(G) + 1, \text{ and the distance } d = \deg(G) + 2,$$

since the degree of the divisor of any differential is equal to

$$\deg(\operatorname{div}(\omega)) = -2$$

so there are 2 more poles than zeros.

We obtain the relationship

$$d = r + 1, \quad n = q + 1.$$

Along with the mapping ϕ, we now consider the mapping

$$L(G) \xrightarrow{\phi^*} A^n$$

$$\phi^* : f \to (f(P_1), f(P_2), ..., f(P_n)).$$

Every function f of $L(G)$ has no poles at the points $\{P_1, ..., P_n\}$, hence its values $f(P_i)$ are defined. The kernel of the mapping ϕ^* consists of the functions taking on zero at each of the points $P_1, ..., P_n$. Therefore, the kernel is zero for $\deg(G) < n$. All the functions of $L(G)$ are of the form

$$f = \frac{\psi(z)}{g(z)},$$

where $\psi(z)$ is an arbitrary polynomial of degree no greater than $r - 1 = \deg g(z)$. Therefore, $L(G)$ is generated by the functions

$$\{1, \frac{1}{g(z)}, \frac{z}{g(z)}, \ldots, \frac{z^{r-2}}{g(z)}\}. \tag{1.26}$$

Thus ϕ^* defines an embedding of P^1 into a normal rational curve with the parametrization

$$x_0 : x_1 : \ldots : x_{r-1} = g(z) : 1 : z \ldots \cdot z^{r-2}.$$

One can also choose another basis of $L(G)$ by decomposing $\psi(z)/g(z)$ into partial fractions

$$\{1, \frac{1}{z - \beta_1}, \frac{1}{(z - \beta_1)^2}, \ldots, \frac{1}{(z - \beta_1)^{m_1}}, \frac{1}{(z - \beta_2)}, \ldots, \}. \tag{1.27}$$

The two bases define the same error-checking matrix of the normal rational code. We show that the corresponding code coincides with the code $U = \Omega(G - D)$.

Indeed, for any $\omega \in (G - D), f \in L(G)$

$$\sum \text{Res}_{p_i}(\omega) \ f(P_i) = \sum \text{Res}_{p_i}(\omega f) = 0,$$

(the differential ω has poles only at the points P_i, on multiplying it by f we obtain a new differential ωf, whose poles coincide with the poles of ω since f has a pole of multiplicity no greater than m_j at the point β_j, ω has a zero of multiplicity no less than m_j at this point).

It is convenient to represent the error-checking matrix in polynomial form. The vector code $x = (a_1, a_2, \ldots, a_n)$ is identified with the differential

$$dz \sum \frac{a_i}{z - \alpha_i} \equiv 0 (\text{mod} \, g(z)).$$

In the polynomial ring, the element $1/(z - \alpha_i)$ modulo $g(z)$:

$$\frac{1}{z - \alpha_i} \equiv \frac{g^{-1}(\alpha_i)g(z) - 1}{z - \alpha_i} (\text{mod} \, g(z)),$$

therefore, the error-checking matrix can be represented as a row of polynomials

$$\left(\frac{g^{-1}(\alpha_1)g(z)-1}{z-\alpha_1},\dots,\frac{g^{-1}(\alpha_{n-1})g(z)-1}{z-\alpha_{n-1}}\right). \tag{1.28}$$

Such a representation is very convenient for decoding.

Let $q = 2^m$. We single out of $\Omega(C-D)$ all the differentiable whose residues lie in F_2. We obtain a binary code whose error-checking matrix results from decomposing a binary code in (1.28) into a basis of the field F_{2^m} over the field F_2. In this case the number of control symbols satisfies the inequality

$$r' \leqslant mt + 1, \quad \deg g(z) = 2t.$$

To estimate the code separation, we consider the Hamming sphere V_{d-1} of radius $d-1$ in A^n. This sphere is identified with the set of differentials of $\Omega(G-D)$ which have no greater than $d-1$ poles. Each of these differentials can pass through no more than $(d-3)/s$ polynomials of degree s irreducible over F_{2^m}. Therefore, if

$$|V_{d-1}|\frac{(d-3)}{s} < N(s) \tag{1.29}$$

where $N(s)$ is the number of polynomials of degree s irreducible over F_{2^m}, then there exists a polynomial $g(z)$ generating a code which does not intersect V_{d-1} and, consequently, has the code separation no less than d. The number $N(s)$ is known to be expressible through the Möbius function

$$N(s) = \frac{1}{s}\sum_{d/s}\mu(d)q^{s/d}. \tag{1.30}$$

The Möbius function $\mu(n)$ is defined for all integers n in the following manner

$$\mu(n) = \begin{cases} 0 & \text{if } p_i^2 \quad n \text{ for a } p_i \\ (-1)^k & \text{if } n = p_1\ p_2...p_k, \text{i.e.,} n \text{ is free of squares} \\ 1 & \text{if } n = 1 \end{cases}$$

This function comes up most frequently in connection with the solution of the following problem of 'the conversion of an arithmetic function. Given the values of the sums

$$F(n) = \sum_{d \,/\, n} f(d) \tag{1.31}$$

of an arithmetic function $f(n)$ over all divisors of n, one has to find the value of the function $f(n)$. The answer is given by the following Möbius inversion formula

$$f(n) = \sum_{d \,/\, n} \mu(d) F(n \,/\, d) \tag{1.32}$$

To prove this fact it suffices to substitute (1.31) into (1.32): $g_s(z)$ of degree s is a divisor of the polynomial $z^{q^m} - z$, where $s \,/\, m$, so

$$z^{q^m} - z = \prod_{s \,|\, m} \prod_{l_r} g_s(z),$$

where the second product is taken over all irreducible polynomials of degree s with unit leading coefficient. On equating the degrees, we obtain

$$q^m = \sum_{s \,|\, m} s N(s). \tag{1.33}$$

On applying the inversion formula (1.32), we obtain (1.30). Inequality (1.29) usually yields a more accurate estimate of the code separation than that based on the degree of the divisor G, in particular for large n.

11. Code Projections

Let $y_0, \ldots, y_{r-1}$ be coordinates in the space P^{r-1}. Projecting the columns Q_j of the error-checking matrix from the point P onto the hyperplane S described by the equation $y_{r-1} = 0$ we obtain a new code whose columns are of the form

$$\tilde{Q}_j = Q_j - \frac{S(Q_j)}{S(P)} P : \begin{cases} \tilde{Q}_j = \lambda P + \mu Q_j \\ 0 = S(\tilde{Q}_j) = \lambda S(P) + \mu S(Q_j) \end{cases}$$

(the point P is assumed not to lie in the plane S, so $S(P) \neq 0$, where $S(X)$ is the last coordinate of X).

The resulting code is called the projection of the initial code. Let us see how the code parameters change under the projection. First it is clear that the number of control symbols reduces at least by 1. If two points Q_1 and Q_2 are projected to the same point $\tilde{Q}$, then the points Q_1, Q_2, and P are clearly collinear and, conversely, if $P \neq \lambda Q_i + \mu Q_j$ for any values of i, j, λ, and μ, then the length of the code is preserved. The set of all linear combinations $\{\lambda_{i_1} Q_{i_1} + \lambda_{i_2} Q_{i_2} + ... + \lambda_{i_m} Q_{i_m}\}$ is called the m-hull of the code. Thus the code length is preserved, if $P \notin$ 2-hull of the code.

Let the initial code be of weight d. Then there exists a linear combination

$$Q = \sum_{j=1}^{d-1} b_j Q_j$$

such that

$$\tilde{Q} = \sum_{j=1}^{d-1} b_j \tilde{Q}_j.$$

Thus the weight does not increase under projection. Conversely,

$$\tilde{Q} - \sum_{j=1}^{d-1} b_j \tilde{Q}_j$$

can be zero for nonzero value of

$$Q - \sum_{j=1}^{d-1} b_j Q_j = cP.$$

Thus the code weight is preserved under projection if the centre of projection P does not belong to the d-hull of the code.

By applying this fact, one may improve the codes projecting the codes from a point. The d-hull of the code can be interpreted in the following manner: the hull consists of all syndromes corresponding to the vectors of the Hamming sphere of radius d. In other words, the hull results from multiplication of all vectors of V_d by the error-checking matrix. It follows that the size of the d-hull is no greater than $|V_d|$. Whence the following lower bound for the linear code is valid

$$r \leqslant \log_q |V_d|$$

(the Varshamov-Hilber bound). For an arbitrary code it is difficult to decide without complete enumeration whether a point of the space P^{r-1} belongs to the *d*-hull, however, the suggested method of 'contracting' the matrix remains valid for zero characteristic, say, for integral matrices in which case the desired centre of projection can readily be found from the maximum modulus of a number.

Let

$$AX = b, \quad X = (x_1, ..., x_n)$$ (1.34)

$$x_j \in R_j, \quad R_j = [0, 1, ..., d_j]$$

be a set of linear integer equations. In terms of linear programming any vector satisfying (1.34) is called a plan. We now project the columns of A and the vector b from a point P onto the hyperplane S. Clearly, every plan of the initial problem is a plan of the resulting problem. Conversely, let $\sum \tilde{Q}_j x_j = \tilde{b}$ be a plan of the problem resulting from projection. Then

$$\sum Q_j x_j - b = \frac{P}{S(P)} S(\sum Q_j x_j - b)$$

The vector

$$Y = \frac{\sum Q_j x_j - b}{S(\sum Q_j x_j - b)}, \quad x_j \in R_j \quad j = 1, ..., n$$

is said to be a normalized residual. The set of plans is preserved under projection provided the centre of projection does not belong to the set of such vectors. On performing a number of iterations, we obtain a single linear condition with the same set of planes as of the initial problem (Knapsack problem).

Maximize the function

$$C = 10x_1 + 8x_2 + 6x_3 + 7x_5 + 7x_5 + 6x_6 + 9x_7 + 7x_8 + 8x_9 +$$

$$+ 8x_{10} + 8x_{11} + 9x_{12}$$

subject to

$$6x_1 + 2x_2 + 7x_3 + 8x_4 + 9x_5 + 8x_6 + 6x_7 + 7x_8 + 9x_9 + 7x_{10} +$$

$$+ 2x_{11} + 10x_{12} \leqslant 10$$

$$4x_1 + 6x_2 + 6x_3 + 7x_4 + 7x_5 + -x_6 + 5x_7 + 5x_8 + 7x_9 + 9x_{10} +$$

$$+ 4x_{11} + 10x_{12} \leqslant 15$$

$$9x_1 + 4x_2 + 9x_3 + 6x_4 + 8x_5 + 10x_6 + 3x_7 + 3x_8 + 9x_9 + 10x_{10} +$$

$$+ 7x_{11} + 11x_{12} \leqslant 14$$

$$x_i = \{0,1\}, \quad i = 1,2,...,12$$

Applying consecutive projection, we reduce this problem to the equivalent knapsack problem, i.e., a problem with a single constraint.

Introducing additional variables

$$0 \leqslant x_{13} \leqslant 10, \quad 0 \leqslant x_{14} \leqslant 15, \quad 0 \leqslant x_{15} \leqslant 14.$$

We reduce the initial set of inequalities to the set of equations with the matrix A:

x_1	x_2	x_3	x_4	x_5	x_6	x_7	x_8	x_9	x_{10}	x_{11}	x_{12}	x_{13}	x_{14}	x_{15}	b	P
6	2	7	8	9	8	6	7	9	7	2	10	1			10	-100
4	6	6	7	7	9	5	5	7	9	4	10		1		15	0
9	4	9	6	8	10	3	3	9	10	7	11			1	14	10

The centre of projection P is chosen in the following way. In the residual vector

$$Y = \frac{\sum Q_j x_j - b}{S(\sum Q_j x_j - b)}$$

the first component takes on values in the segment $[81,81]$, since $(\sum Q_j x_j - b)$ varies in the range $[-10,81]$, and the minimum absolute value of $S(\sum Q_j x_j - b)$ is 1.

Therefore, if we take, for example, $P = (-100,0,1)$, then P definitely cannot be a residual vector, hence projection produces no additional feasible solutions. Here we showed the simplest way to choose P. More through analysis of the initial matrix enables us to choose the centre of projection with smaller coordinates.

After projection on the lower row of A, we obtain a new matrix

x_1	x_2	x_3	x_4	x_5	x_6	x_7	x_8	x_9	x_{10}	x_{11}	x_{12}	x_{13}	b	P
906	402	907	608	809	1008	306	307	909	1007	702	1110	1410		
4	6	6	7	7	9	5	5	7	9	4	10	15		

The last projection leads to the single row

x_1	x_2	x_3	x_4	x_5	x_6	x_7	x_8	x_9	x_{10}
90604	40206	90706	60807	80907	100809	30605	30705	90907	100709

.

Thus we come to the following 0-1 knapsack problem:

$$\max C = 10x_1 + bx_2 + \ldots \rightarrow \max,$$

subject to

$$90604x_1 + 40206x_2 + \ldots + 111010x_{12} \leqslant 141015,$$

$$x_i = \{0,1\}, \quad -i = 1,\ldots,12.$$

This problem can be solved by a pseudo-polynomial algorithm with complexity $0(n^2,C)$ (see C. Papadimitriou, K. Steiglitz, Combinatorial Optimization, Moscow, Mir Publishers, 1985).

This algorithm consists of consecutive generation of numerical arrays. An element of each array is a pair (S,c), where S is a set of variables $S = \{x_{i_1}, x_{i_2}, \ldots, x_{i_k}\}$ and c is the corresponding cost (the value of the objective function).

We begin with the empty set:

$$M_0 = \{(\varnothing,0)\}.$$

Then the variable x_1, is added

$$M_1 = \{(\varnothing,0),(\{1\},10)\},$$

after checking the condition $90604 \leqslant 141015$.

We further check, whether x_2 is feasible and repeat the check for the pair $\{x_1,x_2\}$. The corresponding elements of the array should be ordered in costs

$$M_2 + \{(\varnothing,0), (\{2\},8), (\{1,2\}, 18)\}.$$

If two elements of the array prove to have the same cost, then only one of them is saved, the one which has the smallest sum of the corresponding weigths (the number 90604, 40206, . ..)

$$M_3 = \{(\varnothing,0),(\{3\},6),8),(\{1\},10),(\{2,3\},14),$$

$$(\{1,2\},18)\}.$$

$$M_4 = \{(\varnothing,0),(\{3\},6),(\{4\},7),(\{2\}8),(\{1\},10),$$

$$(\{2,3\},14),(\{2,4\},15),(\{1,2\},18)\}.$$

$$M_5 = \{(\varnothing,0),(\{3\},6),(\{4\},7),(\{2\}8),(\{1\},10),$$

$$(\{2,3\},14),(\{2,4\},15),(\{1,2\},18)\}.$$

After generating the array M_5 we see that $(\{2,4\},15)$ and $(\{2,5\},15)$ coincide, however, since

$$40206 + 60807 < 40206 +'80907,$$

only the first element is left.

$$M_6 = \{(\varnothing,0),(\{3\},6),(\{4\},7),(\{2\},8),(\{1\},10),$$

$$(\{2,3\},14),(\{2,4\},15),(\{1,2\},18)\}.$$

$$M_7 = \{(\varnothing,0),(\{3\},6),(\{4\},7),(\{2\},8),(\{7\},9),$$

$$(\{1\},10),(\{2,3\},14),(\{2,4\},15)(\{4,7\},16),$$

$$(\{2,7\},17),(\{1,2\}18),(\{1,7\},19),(\{2,4,7\},24)\}.$$

$$M_8 = \{(\varnothing,0),\{3\},6),(\{8\},7),(\{2\},8),(\{7\}9),$$

$$(\{1\},10),(\{3,8\},13),(\{4,8\},14),(\{2,8\},15),$$

$$(\{7,8\},16),(\{2,7\},17),(\{1,2\},18),(\{1,7\},19),$$

$$(\{2,4,8\},22),(\{4,7,8\},23),(\{2,7,8\},24)\}.$$

$$M_9 = M_8$$

$$M_{10} = M_9$$

$$M_{11} = \{(\varnothing,0),(\{3\},6),(\{8\},7),(\{2\},8),(\{7\},9),$$

$$(\{1\},10),(\{3,8\},13),(\{4,8\},14),(\{2,8\},15),$$

$$(\{7,8\},16),(\{2,7\},17),(\{1,2\},18),(\{1,7\},19),$$

$$(\{2,4,8\},22),(\{4,7,8\},23),(\{2,7,8\},24),$$

$$(\{2,7,1\},25)\}.$$

$$M_{12} = M_{11}.$$

Thus the optimal solution of the problem is $\{2,7,11\}$ with the cost $c = 25$.

12. Linear Hashing

In various computer-aided information systems, one normally has to find an address by a key. The key is a set $k_1, k_2, ..., k_N$ of various n-dimensional binary vectors, while the address is a set $A_1, A_2, ..., A_N$ of Tm-dimensional binary vectors $(m < n)$. The simplest hashing method consists of multiplication of a key by a fixed $m \times n$ matrix T. In this case it is needed that

$$Tk_i = A_i \neq A_j = Tk_j$$

for all $i \neq j$ (perfect hashing).

Consecutive projection enables us to find the matrix T for a given set of keys.

Suppose that the matrix T defines a perfect hash-function. On projecting this matrix from a point P, we obtain the perfect hash-function $\tilde{T}$ if and only if

$$P \neq T(k_i - k_j), \quad i,j = 1,...,N, \quad i \neq j.$$

It is convenient to start projection with the unit $n \times n$ matrix, and to reduce the number of rows by 1 at every step.

Example: $n = 7$, $N = 6$.

K_1	K_2	K_3	K_4	K_5	K_6
1	1	0	1	0	1
0	0	1	0	0	0
1	0	1	0	1	0
1	1	0	1	0	1
0	0	1	1	1	0
0	0	0	0	0	1
0	0	1	1	1	1

	$-\frac{k_1}{k_2}$	$-\frac{k_1}{k_3}$	$-\frac{k_1}{k_4}$	$-\frac{k_1}{k_5}$	$-\frac{k_1}{k_6}$	$-\frac{k_2}{k_3}$	$-\frac{k_2}{k_4}$	$-\frac{k_2}{k_5}$	$-\frac{k_2}{k_6}$	$-\frac{k_3}{k_4}$	$-\frac{k_3}{k_5}$	$-\frac{k_3}{k_6}$
1	0	1	0	1	0	1	0	1	0	1	0	1
1	0	1	0	0	0	1	0	0	0	1	1	1
1	1	0	1	0	1	1	0	1	0	1	0	1
1	0	1	0	1	0	1	0	1	0	1	0	1
1	0	1	1	1	0	1	1	1	0	0	0	1
1	0	0	0	0	1	0	0	0	1	0	0	1
1	0	1	1	1	1	1	1	1	1	0	0	0

The points P_1, P_2, and P_3 do not belong to the array $S = \{k_i - k_j\}$, therefore they can be chosen as centres of projection. As a result we have

																P
0100000	0	1	0	0	0	1	0	0	0	1	1	1	0	0	0	1
0010000	1	0	1	0	1	1	0	1	0	1	0	1	1	0	1	0
0000100	0	1	1	1	0	1	1	1	0	0	0	1	0	1	1	1
0000010	0	0	0	0	1	0	0	0	1	0	0	1	0	1	1	1

The vector $P = (1,0,1,1)$ does not belong to the array $\tilde{S}$. After projection from this point on the lower row we obtain

0100010	0	1	0	0	1	1	0	0	1	1	1	0	0	1	1
0010000	1	0	1	0	1	1	0	1	0	1	0	1	1	0	1
0000110	0	1	1	1	1	1	1	1	1	0	0	0	0	0	0

Now the array S contains all the vectors and we can no longer proceed with projection. The resulting matrix

$$T = \begin{bmatrix} 0 & 1 & 0 & 0 & 0 & 1 & 0 \\ 0 & 0 & 1 & 0 & 0 & 0 & 0 \\ 0 & 0 & 0 & 0 & 1 & 1 & 0 \end{bmatrix}.$$

defines the perfect hash-function for the given set of keys:

$$A_1 = Tk_1 = \begin{bmatrix} 0 \\ 1 \\ 0 \end{bmatrix},$$

$$A_2 = Tk_2 = \begin{bmatrix} 0 \\ 0 \\ 0 \end{bmatrix},$$

$$A_3 = \begin{bmatrix} 1 \\ 1 \\ 1 \end{bmatrix} \quad A_4 = \begin{bmatrix} 0 \\ 0 \\ 1 \end{bmatrix} \quad A_5 = \begin{bmatrix} 0 \\ 1 \\ 1 \end{bmatrix} \quad A_6 = \begin{bmatrix} 1 \\ 0 \\ 1 \end{bmatrix}.$$

The complexity of the described algorithm of constructing the matrix T is $O(N^2)$.

Comments

The notion of an correcting code was introduced by Hamming [4]. He developed codes correcting a single error and established the upper bound for codes. The codes described in this chapter was first presented by Hocquenhem [5]. The properties of ovals, twisted cubic, and normal rational curve are taken from Segre [9], [10].

In my presentation of the geometric decoding method, which reduces to finding the intersection points of a spatial curve with a linear manifold, I follow the book of Semple and Kneebon [11].

The most complete bibliography on coding theory is presented in the book by F.J. MacWilliams and N.J.A. Sloane, The theory of error-correcting codes, I, II, North-Holland, 1977.

CHAPTER 2

Decoding and Rational Approximations

1. Continued Fractions

In the previous chapter we introduced normal rational codes and developed a method of decoding based on a simple geometric idea. The code vectors were described algebraically in terms of differential representation. Such a representation leads to a new point of view of the decoding.

We make use of the error-checking matrix (1.28). In this case the syndrome $S(z)$ is a polynomial resulting from the multiplication of a row of polynomials (1.28) by the code vector. The decoding problem is as follows: find a rational function $\psi(z)/f(z)$ of the minimum degree such that

$$\psi/f - S \equiv 0 (\mathrm{mod}\, g(z)). \tag{2.1}$$

This problem is equivalent to the problem which Huygens came across in the 17th century as constructing a model of the solar system using great-wheels, the problem which led him to the discovery of one of the most abstract area of modern mathematics - the Diophantine approximations.

47

The model of the solar system which was constructed by Huygens consisted of gear-wheels with a given ratio of angular velocities. Since the angular veloci-ties are inversely proportional to the number of teeth, it was technologically difficult to manufacture gears with a great number of teeth. The following mathematical problem arose. For a given real number α, find a rational fraction with the smallest possible denominator, which approximations α to the prescribed accuracy

$$|a/b - \alpha| \leqslant \epsilon.$$

If any other fraction c/d approximating α to the same accuracy has a greater denominator than a/b, then a/b is called the best rational approximation for α.

To find such fractions, the continued fractions

$$\alpha = q_0 + \cfrac{1}{q_1 + \cdots}, \qquad (2.2)$$

are used.

The continued expansion is obtained by the Euclidean algorithm if α is rational (in which case we obtain a finite fraction) or by taking the greatest integer not exceeding if α is real (the infinite continued fraction).

For any continued fraction

$$\alpha = [q_0, q_1, \ldots]. \qquad (2.3)$$

the fraction P_k/Q_k, which is a convergent of the continued fraction, is defined as

$$\begin{cases} P_{-1} = 1 \\ Q_{-1} = 0 \end{cases}, \begin{cases} P_0 = q_0 \\ Q_0 = 1 \end{cases}, \begin{cases} P_k = q_k P_{k-1} + P_{k-2} \\ Q_k = q_k Q_{k-1} + Q_{k-2} \end{cases} k = 1, 2, \ldots, \quad (2.4)$$

which is equivalent to the definition

$$P_0/Q_0 = q_0, \ P_1/Q_1 = [q_0, q_1], \ldots, P_k/Q_k = [q_0, \ldots, q_k]. \qquad (2.5)$$

There is the identity

$$\frac{P_k}{Q_k} - \frac{P_{k-1}}{Q_{k-1}} = \frac{(-1)^k}{Q_k Q_{k-1}}. \qquad (2.6)$$

It follows that the fractions convergent to a continued fraction are irreducible: $(P_k, Q_k) = 1$ and P_k / Q_k is the best approximation for α.

The approximation accuracy is estimated as

$$\left| \alpha - \frac{P_k}{Q_k} \right| \leqslant \frac{1}{Q_k Q_{k+1}}. \tag{2.7}$$

Thus for any real α and $\tau \geqslant 1$ there exists an irreducible fraction a / b such that

$$|\alpha - a / b| < 1 / b\tau, \quad 0 < b \leqslant \tau. \tag{2.8}$$

(The Dirichlet theorem).

The computations are carried out according to the following scheme

q_k		q_0	q_1	...	q_n
P_k	$P_{-1} = 1$	$P_0 = q_0$	P_1		P_n
Q_k	$Q_{-1} = 0$	$Q_0 = 1$	Q_1		Q_n

For example, the number $43/11$ has the following expansion into a continued fraction: $43/11 = (3,1,10)$

q_k	3	1	10
P_k	1	3	4	43
Q_k	0	1	1	11

2. The Fermat Theorem on the Sum of Two squares.

Let p is a prime number, $\tau = \sqrt{p}$, and let k be an integer. We set $\alpha = k / p$. Then it follows from (2.8) that

$$\left| \frac{k}{p} - \frac{z}{x} \right| < \frac{1}{x\sqrt{p}}, \quad 0 < x \leqslant \sqrt{p}, \tag{2.9}$$

where z, x are integers. Let

$$y = kx - zp,$$ (2.10)

then

$$y \equiv kx(\mathrm{mod}\, p),$$ (2.11)

$$y^2 \equiv k^2 x^2(\mathrm{mod}\, p).$$ (2.12)

The equation $k^2 = -1(\mathrm{mod}\, p)$ is either solvable, or insolvable depending on whether p is $4n + 1$ or $4n + 3$. In the former case we have

$$y^2 + x^2 \equiv 0(\mathrm{mod}\, p)$$

and since $|y| < \sqrt{p}$ and $|x| \leqslant \sqrt{p}$, it follows that

$$y^2 + x^2 = p.$$ (2.13)

Therefore, equation (2.13) with $p = 4n + 1$ is always solvable in relatively prime x, y (the Fermat theorem).

3. Approximations in the *p*-Adic Metric

It follows from (2.9), (2.10) and (2.11) that there exists a rational number y / x such that

$$y \equiv kx(\mathrm{mod}\, p),$$ (2.14)

$$\mathrm{Max}\{|y|, |x|\} \leqslant \sqrt{p}.$$

Here k is an arbitrary integer and it suffices to know only the residue of this number modulo p. The congruence in (2.14) can be written in the form

$$\|y / x - k\|_p \leqslant \epsilon, \quad \epsilon = p^{-1}.$$

The same is true if p is replaced by its power p^m:

$$\|y / x - k\|_p \leqslant \epsilon, \quad \epsilon = p^{-m},$$ (2.15)

$$\text{Max}\{\,|y|\,,|x|\,\} < \sqrt{q^m} = \frac{1}{\sqrt{\epsilon}}.$$

Thus (2.15) characterizes a rational approximation in the p-adic metric and is a p-adic counterpart of the Dirichlet theorem.

One can take an arbitrary polynomial $S(z)$ instead of k and consider approximation in the norm $\|\alpha\|_{z-\beta}$. Then (2.15) takes the form

$$\|\psi(z)/f(z) - S(z)\|_{z-\beta} \leqslant \epsilon, \quad \epsilon = \theta^m, \tag{2.16}$$

$$\text{Max}\{\|\psi(z)\|, \ \|f(z)\|\} \leqslant \frac{1}{\sqrt{\epsilon}}.$$

which is equivalent to the congruence

$$\psi(z) - f(z)S(z) \equiv 0(\text{mod}\,(z-\beta)^m),$$

$$\text{Max}\{\deg\psi, \deg f\} \leqslant m/2.$$

. Finally, considering jointly approximations in the norms $\|\alpha\|_{z-\beta_1}$, $\|\alpha\|_{z-\beta_2}$, $g(z) = (z-\beta_1)^{m_1}...(z-\beta_l)^{m_l}$, we obtain a g-adic analog of the Dirichlet theorem:

For any polynomials $g(z)$ and $S(z)$ there exists a pair of polynomials ψ and f such that

$$\psi(z) - f(z)S(z) \equiv 0(\text{mod}\,g(z))$$

$$\deg f \leqslant t, \ \deg\psi < t \tag{2.17}$$

$$\deg g(z) = 2t$$

This pair is unique: if ψ_1/f_1 and ψ_2/f_2 satisfy (2.17), then

$$\psi_1/f_1 \equiv \psi_2/f_2(\text{mod}\,g(z))$$

$$\psi_1/f_1 - \psi_2/f_2 \equiv 0(\text{mod}\,g(z))$$

which is impossible since $\deg\psi_1 f_2 < 2t$, $\deg\psi_2 f_1 < 2t$.

From (2.9), (2.10), and (2.11) it is clear how to find a rational function ψ/f satisfying (2.17):

$$\psi(z) = S(z)f(z) - g(z)\tilde{\psi}(z)$$

and $\tilde{\psi}/f$ is a convergent of S/g.

4. Decoding Algorithm

We now state a decoding algorithm based on rational approximation to the syndrome $S(z)$ or, which is the same, a method of solving congruence (2.17) with small degrees of the polynomials ψ and f.

Let deg$g = 2t$. Then the code corrects t errors.

(1) Evaluate the syndrome

$$S(z) = \sum a_i \frac{g^{-1}(\alpha_i)g(z)-1}{z-\alpha_i}.$$

Here $\{a_i\}$ are the components of the vector output.

(2) Expand $S(z)/g(z)$ into a continued fraction simultaneously computing the convergents P_k/Q_k of the fraction. Stop the computation, once degQ_k is greater than t.

The polynomial $f = Q_{k-1}$ is an error locator.

3. Compute $\psi = Q_{k-1}S - gP_{k-1}$. The fraction ψ/f is that we desired to find.

4. Break ψ/f into irreducible fractions:

$$\psi/f = \sum_{j=1}^{t} \frac{b_i}{z-\alpha_i}.$$

The pole α_i points out the position of an error and the residue b_i is equal to the magnitude of distortion.

Example. $q = 2^5$, $g(z) = (z^3+z+1)^2$, the errors occur in the positions corresponding to 1 and 0, i.e., the error locator is of the form $(z+1)z$.

$$S(z) = (z^5+z) + (z^5+z^4+z^3+z^2) = z^4+z^3+z^2+z$$

$$S/g = [0,z^2+z,z^4+z^3+z^2+z]$$

q_k			0	z^2+z	$z^4+z^3+z^2+z$
P_k	1	0	1	$z^4+z^3+z^2+z$	
Q_k	0	1	z^2+z	z^6+z^2+1	

$$f = Q_2 = (z+1)z,$$

$$\psi = Q_2 S - P_2 g = (z^2+z)(z^4+z^3+z^2+z) - (z^6+z^2+1) = 1.$$

Consider the congruence

$$f(z)S(z) - \psi(z) \equiv 0 \pmod{z^{2t}}.$$

Let

$$S(z) = s_0 + s_1 z + \dots + s_{2t-1} z^{2t-1},$$

$$f(z) = a_0 + a_1 z + \dots + a_t z^t,$$

$$a_0 \neq 0, \quad \deg \psi(z) < t.$$

Then

$$a_t s_0 + a_{t-1} s_1 + \dots + a_0 s_t = 0$$

$$a_t s_t + a_{t-1} s_2 + \dots + a_0 s_{t+1} = 0$$

$$\dots\dots\dots\dots\dots\dots\dots\dots\dots \qquad (2.18)$$

$$a_t s_{t-1} + a_{t-1} s_t + \dots + a_0 s_{2t-1} = 0.$$

The matrix

$$\begin{pmatrix} s_0 & s_1 & \dots & s_t \\ s_1 & s_2 & \dots & s_{t+1} \\ \dots\dots\dots\dots\dots\dots \\ s_{t-1} & s_t & \dots & s_{2t-1} \end{pmatrix},$$

every row of which is the previous one shifted to the left is called Hankel. Similarly, a matrix

$$\begin{pmatrix} s_t & s_{t-1} & \cdots & s_0 \\ s_{t-1} & s_t & \cdots & s_1 \\ & & \cdots\cdots\cdots & \\ s_{2t-1} & s_{2t-2} & \cdots & s_{t-1} \end{pmatrix}$$

every row of which is the previous one shifted to the right is called Toeplitz.

The geometric method of decoding described in Chapter 1 reduces to the solution of system (2.18) for finding the coefficients of the error locator.

Thus the method of decoding based on the expansion into a continued fraction is merely an efficient method of solving Hankel (or Toeplitz) system: a linear system with such a matrix is replaced by a Diophantine equation.

5. Linear Recurrent Sequences and Shift Registers

The ring $K[[z]]$ of power series contains a subring $K[z]$ of polynomials. Although the only invertible elements of the ring $K[z]$ are those of the field K, all the polynomials with nonzero absolute terms have inverse elements in the ring $K[[z]]$. Thus every rational function $\zeta(z) = \psi/f$ with $f(0) \neq 0$ can be associated with a formal series

$$\delta: \zeta(z) = \psi(z)/f(z) = c_0 + c_1 z + \ldots$$

and this is a one-to-one mapping provided that $\zeta(z)$ is irreducible.

The set of all fractions $\zeta(z)$ with nonzero denominator at 0 make up a subring $Q_0(z)$ of the field $K(z)$ of rational functions. This is a local ring, i.e., all irreversible elements of this ring make up an ideal which is the only maximum ideal of the ring $Q_0(z)$.

Let $\zeta(z)$ be a proper fraction, i.e., $\deg\psi < \deg f$,

$$\psi(z) = u_0 + a_1 z + \ldots a_m z^m,$$

$$f(z) = b_0 + b_1 z + \ldots b_n z, \; m < n.$$

The mapping δ shows that $\{a_i\}, \{b_i\} \{c_i\}$ are interrelated as follows

$$
\left.\begin{array}{r}
b_0 c_0 = a_0 \\
b_0 c_1 + b_1 c_0 = a_1 \\
\cdots\cdots\cdots \\
b_0 c_l + \ldots b_{l-1} c_1 + b_l c_0 = a_l
\end{array}\right\} \ l < n,
\tag{2.19}
$$

$$
c_k = -b_0^{-1}(b_1 c_{k-1} + b_2 c_{k-2} + \ldots + b_n c_{k-n}),
\tag{2.20}
$$

$$
k = n, \, n+1, \ldots .
$$

It follows from (2.20) that the coefficients c_i of the series $S(z)$ corresponding to the function ψ / f are the output of the following device called a shift register with a feedback or merely a shift register (Figure 2).

The initial state $(c_0, c_1, \ldots, c_{n-1})$ of the register is defined by the triangular set of equations (2.19).

The sequence $\{c_k\}$, whose members are found from the recurrent formula (2.20), is called a linear recurrent sequence.

All these sequences are generated by shift registers and, conversely any sequence generated by a shift register is linear recurrent.

Let $\{c_k\} = x$ be an arbitrary sequence of elements of the field K. The power series $X(z) = \Sigma_{k \geqslant 0} c_k z^k$ is usually called the z-transformation of sequence x.

One can see that if $X(z)$ can be expressed as a rational function of the ring $Q_0(z)$, then the sequence x is generated by a shift register. Conversely, if a sequence is generated by a shift register, then it satisfies system (2.20) in which the coefficients b_i are defined by the register feedback. Having first n members $(c_0, c_1, c_2, \ldots, c_{n-1})$ of the sequence x, one can uniquely calculate $(a_0, a_1, \ldots, a_m)$ from formulas (2.19).

Therefore, the following general assertion holds true.

Proposition 1. The following three definitions are equivalent.
(a) a sequence is linear recurrent;
(b) a sequence is generated by a shift register with feedback;
(c) the z-transformation of a sequence can be expressed in the form of a proper irreducible fraction, whose denominator is nonzero at the point $z = 0$.

We shall specify the shift register by use a feedback polynomial

$f(z) = b_0 + b_1 z + ... b_n z^n$. On choosing an arbitrary initial state $(c_0, c_1, ..., c_{n-1})$ of this register, we obtain a unique output sequence. The set of such sequences is denoted by $G(f)$, while the number n which is the degree of f, is called the register order.

For an arbitrary sequence $x = \{c_0, c_1, ...,\}$ the sequence $Tx = \{c_1, c_2, ...\}$ is its shift. From the definition we immediately obtain the following.

Proposition 2. The set $G(f)$ is an n-dimensional linear subspace which is invariant under shift.

The sequence $\{c_0, c_1, ...\}$ generated by the shift register is completely defined by its initial elements $s_0 = (c_0, c_1, ..., c_{n-1})$. The latter vector is usually referred to as a register state. The next state $s_1 = (c_1, c_2, ..., c_n)$ of this register can be obtained by multiplying the initial vector by the characteristic matrix P_f of this register;

$$P_f = \begin{bmatrix} 0 & 0 & -b_n / b_0 \\ 1 & 0 & -b_{n-1} / b_0 \\ .... \\ 0 & 0 & -b_1 / b_0 \end{bmatrix},$$

We have

$$s_1 = s_0 P_f, \; s_2 = s_1 P_f, ..., s_n = s_0 P_f^n.$$

A state s is said to be attainable from state s' in the register f, if f switches from s' to s after a number of shifts. This property holds true if and only if $s = s' P_f^m$ from $m \geqslant 0$.

Every state s defines a sequence starting at s. The z-transformation of this sequence is of the form ψ_s / f such that $\deg \psi_s < \deg f$ and these polynomials can have a common multiplier. The polynomials $\psi_s(z)$ is called the s-state polynomial. The state s is uniquely defined by its polynomial $\psi_s(z)$ and can be found

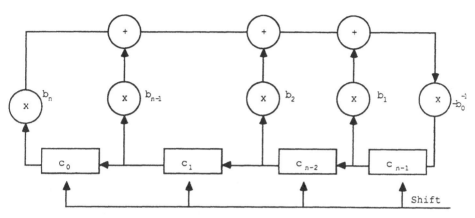

Figure 2.

from system (2.19).

If s is attainable from s' in the register f, then

$$z^m \psi_s(z) - \psi_{s'}(z) \equiv 0 (\mathrm{mod}\, f(z)) \qquad (2.21)$$

from $m \geqslant 0$. The converse is also true.

If ψ_s has no common divisors with f, then this is also true for ψ_s'. If s is attainable from s' exactly in m shifts then ψ_s can be obtained from ψ_s' by the following algorithm

a) expand f / z^m into a continued fraction;

b) find the numerator P_{n-1} of the penultimate convergent of the continued fraction:

c) calculate $(-1)^{n-1} P_{n-1} \psi_{s'}$ and reduce it modulo f; the residue obtained is equal to ψ_s.

For an arbitrary polynomial f of degree n, the dual polynomial f^* is defined as

$$f^*(z) = z^n f(1/z),$$

Clearly, if $f(0) \neq 0$, then $f^*(0) \neq 0$. We define the dual register of f through the feedback polynomial of f^*. If the state s is attainable from s' in the register f, then s' is attainable from s in the register f^*. In other words, f^* runs the same states as f does, but in an inverse order. The polynomial (register) f is said to be periodic, if f divides $z^T - 1$ for an integer $T > 0$. The smallest T is called the period of the polynomial (register).

Proposition 3. The polynomials f and f^* are either both periodic, or both not. Every periodic register generates a periodic sequence regardless of the initial state. All registers over a finite field are periodic.

Proof. If α is a root of f, then α^{-1} is a roof of f^* and the converse is also true. Since the roots of unity make up a multiplicative group, it follows that f and f^* either both divide or both do not divide one and the same polynomial of the form $z^r - 1$. Now let f define a period register with period T. This means that f divides $z^T - 1$, so $z^T - 1 = f\phi$, where ϕ is a polynomial. For any state s we have

$$\psi_s /_f = \psi_s \phi / (z^T - 1) = -\psi_s \phi [1 + z^T + z^{2T}...] \tag{2.22}$$

which proves that the sequence generated by the state s is periodic since $\deg(\psi_s \phi) < T$. Conversely, if a sequence $c_0, c_1,...$ is periodic, then its z-transformation is of the form

$$S(z) = (c_0 + c_1 z +... + c_{T-1} z^{T-1})(1 + z^T + z^{2T}...)$$

for a $T > 0$, hence $S(z) = R(z) / (z^T - 1)$, $\deg R < T$ and after dividing by the common divisor we obtain $S(z) = \phi / f$, where f divides $Z^T - 1$. Finally, if the basic field $\underset{=}{K}$ is finite, then there are a finite number of possible register states. Consequently, after a number of shifts the register comes at the initial state and the sequence cycles. Hence in this case any polynomials is periodic.

When the register is periodic, the set $G(f)$ can be identified whit the set of polynomials $U_s(z)$ which are the initial segments of the power series (within the period T) generated by the state s. It follows from (2.22) that all the polynomials of $G(f)$ are divided by a fixed polynomial $\phi(z) = (z^T - 1) / f(z)$. In this case the set $G(f)$ is invariant under the cyclic shift and is, therefore, called the cyclic code. If (2.21) is multiplied by $I(-\phi(z))$, then we obtain

$$z^m U_s(z) - U_{s'}(z) \equiv 0 (\bmod (z^T - 1)). \qquad (2.23)$$

Since multiplication by z^m followed by reduction modulo $z^T - 1$ is equivalent to a cyclic shift, it follows that (2.23) means that s and s' belong to the same cyclic class. Note that (2.21) is more general since it remains valid for nonperiodic registers. In the case when the ground field $\overline{K}$ is the complex field C, there is a natural notion of convergence and so one can speak about register stability which means that $|c_n| \to 0$ if $n \to \infty$ for any sequence c_n generated by this register. It can be easily shown using the decomposition into partial fractions, that the register is stable if and only if all the roots of f lie outside the unit circle. A polynomial satisfying this condition is also said to be stable. The Rouché theorem leads to a simple stability criterion.

Rouché Theorem (a simplified version). Let f and g be two polynomials such that $|f(z)| > |g(z)|$, $|z| = 1$.

Then the number of zeros of the polynomial $f(z) + g(z)$ in the circle $|z| \leqslant 1$ is equal to the number of zeros of $f(z)$ in the same circle.

Lemma 1. If the polynomial $f(z) = b_0 + ... + b_n z^n$ is stable then $|b_n / b_0| < 1$.

Proof. Let $f(z) = b_n(z - \alpha_1)(z - \alpha_2)...(z - \alpha_n)$. Hence $|b_0| = |b_n \Pi \alpha_i|$, $|b_n / b_0| = \Pi |\alpha_i|^{-1}$, and since for any root $|\alpha_i| > 1$, it follows that $|b_n / b_0| < 1$.

Lemma 2. For any real polynomial $f(z)$ the equality $|f(z)| = |f^*(z)|$ holds for all $|z| = 1$.

Proof. The condition $z = 1$ is equivalent to the fact that $z^{-1} = \overline{z}$, since in this case $z = e^{i\phi}$. Therefore,

$$|f^*(z)| = |z^n f(1/z)| = |z|^n |f(\overline{z})| = |z^n| |\overline{f(z)}|$$

for all $|z| = 1$.

Theorem 3. A real polynomial $f(z) = b_0 + ... + b_n z^n$ is stable if and only if $|b_n / b_0| < 1$ and the polynomial $\psi = f b_0 - b_n f^*$ of a smaller degree is stable.

Proof. Let $|b_n / b_0| < 1$ (otherwise by Lemma 1 the polynomial is not stable). Then

$$|b_0 f| = |b_0| |f| > |b_n| |f^*| = |b_n f^*|$$

for all $|z| = 1$ (Lemma 2). It follows from the Rouché theorem that ψ has as many roots in the circle $|z| \leqslant 1$ as f.

The stability criterion follows from reiteration of Theorem 3 (at every iteration the degree of the polynomial reduces at least by 1).

Example 1. The register is defined by the feedback polynomial

$$f = 4 + 3z + 2z^2 - 2z^3,$$

first iteration.

$$|-2/4| < 1,$$

second iteration.

4	3	2	-2	$\times$	4
-2	2	3	4	$\times$	(-2)
12	16	14			

$$|14/12| > 1.$$

The register is not stable.

Example 2. The Fibonacci sequence is defined by the recurrence relation $c_n = c_{n-1} + c_{n-2}$ with the initial values $c_0 = 0$, $c_1 = 1$. Thus this sequence is 0, 1, 1, 2, 3, 5, 8, 13, . .. In this case the feedback polynomial takes the form $f = 1 - z - z^2$. Since $|b_n| = |b_0|$, it follows that the register is not stable as can be seen straightforwardly.

Example 3. $f = 2 - z + z^2 - z^3$

first iteration:

$$|-1/2| < 1,$$

second iteration:

2	-1	1	-1	$\times$	2
-					
	-1	1	-1	2 $\times$	(-1)
	3	-1	1		

$$|1/3| < 1.$$

third iteration:

3	-1	1	$\times$	3
-				
	1	-1	3 $\times$	1
	8	-2		

$$|2/8| < 1.$$

The register is stable.

The register identification problem is formulated in the following manner. The register is assumed to be a black box, i.e., its structure is unknown and we can only observe its output sequence. Given a finite number of elements of this sequence, one has to determine the register structure, i.e., to find the feedback polynomial. Clearly, this problem is equivalent to the approximation of an arbitrary linear recurrent sequence in the p-adic norm:

$$\psi(z) / f(z) - S(z) \equiv 0 (\bmod z^r)$$

where

$$S(z) = c_0 + c_1 z + \ldots c_{r-1} z^{r-1}$$

corresponds to the initial segment of the output sequence.

Obviously, this problem coincides with the decoding problem for rational codes and the above continued fraction algorithm can readily be applied.

6. Linear Discrete Filters

Let $x[n]$ be a sequence of elements of a field K fed into the input of a device at discrete time moments. The output sequence is denoted by $y[n]$. It is assumed that $x[n] = y[n] = 0$ for $n < 0$. We denote by $X(z)$ and $Y(z)$ the z-transformation of the sequences $x[n]$ and $y[n]$ respectively. Obviously, the z-transformation of the shift $T^m x$ is equal to $z^m X(z)$.

If $x[n]$ and $y[n]$ are interrelated through the difference equation

$$b_0 y[k] + b_1 y[k-1] + ... + b_n y[k-n] = a_0 x[k] + ... + a_m x[k-m]$$

$$k = 0,1,...;a_i, \quad b_j \in K, \quad b_0 b_n \neq 0,$$

then a linear discrete filter (LDF) is said to be given. In case of real or rational field K, we also use the term 'digital filter'

Turning to the z-transformations, we obtain

$$Y(z)(b_0 + b_1 z + ... + b_n z^n) = X(z)(a_0 + ... + a_m z^m).$$

The value

$$\Phi(z) = \frac{Y(z)}{X(z)} = \frac{a_0 + ... + a_m z^m}{b_0 + ... + b_n z^n},$$

is called the transfer function of the filter.

Thus the z-transformation of the LDF output results from the multiplication of the z-transformation of the input by the transfer function.

A sequence $x[n]$ such that $x[0] = 1$ and $x[n] = 0$ for $n > 0$ has the following z-transformation $X(z) = 1$. Such a sequence is called the discrete delta-function. The transfer function is the z-transformation of a sequence $k[n]$ which is the LDF output when the input is the discrete delta-function. The sequence $k[n]$ is called the pulse transfer function of DF.

A filter with $m < n$ is said to be regular. Given two sequences $U[n]$ and

$V[n]$, the new sequence

$$C[n] = \sum_{k=0}^{n} U[k]V[n-k]$$

is called the convolution of U and V. For z-transformations we obtain

$$C(z) = U(z)V(z).$$

The regular linear discrete filter with the transfer function $\Phi(z)$ is now represented in the following form (the first canonical form) (Figure 3). Indeed, $y[n]$ is the convolution between $x[n]$ and $k[n]$. Since $\Phi(z)$ is a proper fraction such that $b_0 \neq 0$, it follows that the sequence $k[n]$ is generated by the shift register corresponding to $\Phi(z) = (a_0 + ... + a_m z^m) / (b_0 + ... + b_n z^n)$. Thus the sequence generated by the shift register is the pulse transfer function of a regular LDF, i.e., describes the natural oscillations of LDF. If the filter is not regular, then $\Phi(z)$ can be expressed as a sum of a proper fraction and a polynomial. In this case a finite number of first members of $k[n]$ are changed.

When the LDF is connected in series, the transfer functions are multiplied. Therefore, any LDF with $\Phi(z) = (a_0 + ... + a_m z^m) / (b_0 + ... + b_n z^n)$ can be represented as two serially connected filters with the transfer functions $1/(b_0 + ... + b_n z^n)$ and $(a_0 + ... + a_m z^m)$ respectively. These filters are called the autoregression and the moving average filters respectively. The representation of an arbitrary LDF as serially connected the autoregression and moving average filters is called the second canonical form of LDF (Figure 3). This form is convenient for digital simulation of filters since in this case it suffices to have memory of size $m + n$. The 1st canonical form is important for theoretical studies since it reduces the theory of LDF to the theory of independent shift register. An LDF is stable if its natural oscillations are attenuated, i.e., if $|k[n]| \to 0$. Since $k[n]$ is generated by the shift register, it follows that the LDF stability is equivalent to that of the corresponding shift register, it depends only on the dominator of the transfer function, and one can use the above-mentioned criterion to study this function. Furthermore, the simplest LDF identification problem is as follows. Assume that the LDF transfer function is unknown. Given a trial input signal, one has to find the filter structure from the filter response, i.e.,

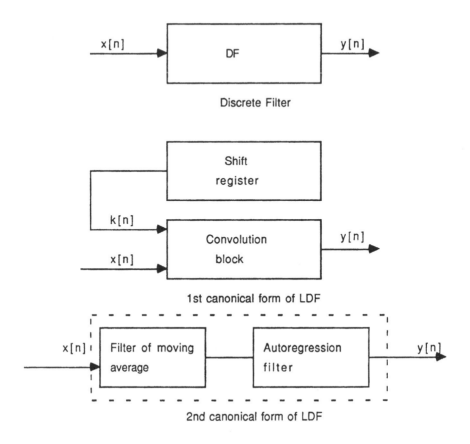

Figure 3.

to find the transfer function. If the discrete delta-function is taken for the trial signal, then the filter response is its pulse transfer function $k[n]$, hence the LDF identification problem reduces to the shift register identification problem and we can apply the continued fraction algorithm.

Example 1. The real stable LDF with the transfer function

$$\phi(z) = (1 + z + z^2)/(2 - z + z^2 - z^3),$$

can be represented as the serially connected moving average filter of the second order and autoregression filter of the third order. We evaluate the filter response to the discrete delta-function

Step	Input	Filter state	Output
1	2	3	4

0	1	00	0	0	0	0
1	2			3		4
1	0	01	1/2	0	0	1/2
2	0	10	3/4	1/2	0	3/4
3	0	00	5/8	3/4	1/2	5/8
4	0	00	3/16	5/8	3/4	3/16
5	0	00	5/32	3/16	5/8	5/32
6	0	00	19/64	5/32	3/16	19/64
7	0	00	21/128	19/64	5/32	21/128

In this case the filter response is its pulse transfer function $k[n]$. This response is shown in Figure 4. This sequence can also be obtained at the output of the shift register (Figure 4) if the initial state is evaluated from the system (2.19). In such a case $a_0 = a_1 = a_2 = 1$, hence we have $c_0 = \dfrac{1}{2}$, $c_1 = \dfrac{3}{4}$, $c_2 = \dfrac{5}{8}$.

Assume that the transfer function is unknown, however, it is known that the sequence $k[n]$ is obtained at the LDF output of order no smaller than 3. Then to obtain a unique structure of the filter it suffices to observe the first 6 values of the output. We associate these values with the polynomial

$$S(z) = \frac{1}{2} + z(3/4) + z^2(5/8) + z^3(3/16) + z^4(5/32) + z^5(19/64)$$

and expand $S(z)/z^6$ into a continued fraction

		0	$z\dfrac{64}{19}-\dfrac{640}{19^2}$	$-z\dfrac{19^3}{2.64^2}+\dfrac{85.19^2}{2.64^2}$	$-z\dfrac{128^2}{2.361^2}-\dfrac{128.512}{2.361^2}$
P_j	1	0	1	$-z\dfrac{19^3}{2.64^2}+\dfrac{85.19^2}{2.64^2}$	$z^2\dfrac{1}{19}-z\dfrac{9}{361}+\dfrac{21}{361}$
Q_j	0	1	$z\dfrac{64}{19}-\dfrac{640}{19^2}$	$-z^2\dfrac{361}{128}+z\dfrac{361.5}{128}-\dfrac{362}{64}$	$z^3\dfrac{64}{361}-z^2\dfrac{64}{361}+z\dfrac{64}{361}-\dfrac{128}{361}$

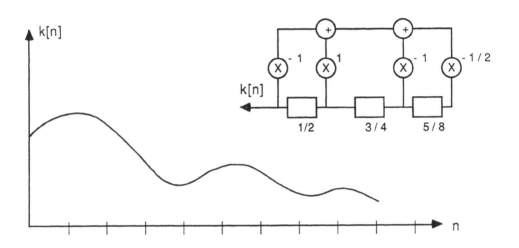

Figure 4.

Evaluating

$$R_3 = Q_3 S - P_3 z^6 = (z^3 - z^2 + z - 2)S(z) -$$

$$- z^6[19/64)z^2 - (9/64)z + 21/64] = -(z^2 + z + 1),$$

we finally obtain

$$\phi(z) = R_3 / Q_3 = (1 + z + z^2)/(2 - z + z^2 - z^3).$$

If the input trial signal were unit stepfunction rather than the delta-function, then the unique identification would require 8 values of the output signal rather than 6, since the z-transformation of the output signal would have the form

$$(1+z+z^2)/(1-z)(2-z+z^2-z^3).$$

Example 2. The LDF identification over the field $GF(2)$ with transfer function

$$\phi(z) = (z+z^2+z^3)/(1+z+z^4).$$

Step	Input	Filter state		Output
0	1	000	0000	0
1	1	001	0000	1
2	1	011	1000	1
3	1	111	1100	0
4	1	111	0110	1
5	1	111	1011	1
6	1	111	1101	1
7	1	111	1110	0
8	1	111	0011	0
9	1	111	0001	0

$$S(z) = z + z^2 + z^4 + z^5 + z^6$$

$$S/z^{10} = [0, z^4 + z^3 + z, z, z^2, z + 1, z]$$

		0	$z^4 + z^3 + z$	z
P_j	1	0	1	z
Q_j	0	1	$z^4 + z^3 + z$	$z^5 + z^4 + z^2 + z$

$$Q = z^5 + z^4 + z^2 + 1$$

$$R = z + z^2 + z^3.$$

Since in this case the z-transformation of the trial signal is $1/(1-z)$, it follows that $\phi(z)$ results from the multiplication of R/Q by $(1-z)$:

$$\phi(z) = (z + z^2 + z^3)/(1 + z + z^4).$$

7. Functions with Rational Spectrum

In previous sections we studied the class G of linear recurrent sequences which is closed under linear operations and multiplication that is the discrete convolution in our context. Thus G is an algebra over the field K. Every element of G can be effectively encoded

a) by use of coefficients $\{a_0,...,a_m,b_0,b_1,...,b_n\}$ of the z-transformation of the sequence $\phi(z)$;

b) by use of the divisor of zero and poles of $\phi(z)$;

c) by use of the initial segment of length $2n$, where n is the order of the sequence (the degree of $\phi(z)$).

In the set of real-valued continuous functions of t there exists a class of functions whose properties are similar to those of linear recurrent sequences.

This is the set G^* of functions such that the Laplace transformation $F(s)$

$$\delta^*: F(s) = \int_0^\infty f(t)e^{-st}dt$$

can be expressed as a proper rational-fractional function.

Denote by $R(s)$ the set of rational-fractional functions of the form

$$F(s) = \frac{\phi(s)}{f(s)}$$

$$\deg \psi < \deg f.$$

The δ^* defines the isomorphism $G^* \rightarrow R(s)$ algebras. Every function of G^*

can be effectively encoded by (a) coefficients $\{a_0,...,a_m,b_0,...,b_n\}$ of the Laplace transformation of this function (b) by the divisors of zeros and poles of this function.

One may conjecture that the functions of G^* enjoy a property similar to property (c) of the linear recurrent sequences. We introduce the 'quantization transformation' T_h which maps every function $f(t) \in G^*$ to a lattice function $f(0), f(h), f(2h),....$ The linear transformation T_h is a mapping of G^* onto G, i.e., any linear recurrent sequence can be obtained by quantization of a continuous function with rational spectrum.

Note that T_h is not a one-to-one correspondence. The kernel G_h^* of this mapping is made up of all the functions $f(t) \in G^*$ which are zero on the lattice $0, h, 2h,....$ Thus under the quantization T_h every linear recurrent sequence of G corresponds to a coset $f(t) + G_h^*$ of the space G^*. The order of this sequence coincides with the smallest order of functions in this coset.

We now assume that the transfer function $\phi(s)$ of a continuous filter is unknown. If a function with rational spectrum is taken as the trial signal, then the output is also a function $F(s)$ with rational spectrum. If the quantization step is arbitrary (random), then it is unlikely that the unknown spectrum of $F(s)$ has a component whose period is divided by h, therefore the spectrum can be uniquely determined by measurement of the output at a finite number of points.

The set $\{f(0), f(h),..., f(2nh)\}$ is put in correspondence with the polynomial $S(z)$, to which the continued fraction algorithm is applied to identify the linear recurrent sequence. We obtain a fraction $\phi(z)$ of degree no greater than n. Further, the poles of $\phi(z)$ are determined which enables us to find the unknown transfer function $\phi^*(s)$ of the continuous filter. If $\alpha_1,...,\alpha_n$ are poles of $\phi^*(s)$, then $e^{-\alpha_1 h}, e^{-\alpha_2 h},..., e^{-\alpha_n h}$ are poles of $\phi(z)$.

The continuous filter $\phi^*(s)$ is stable if all its poles lie in the left half-plane.

The linear-fractional transformations $s = (1-z)/(1+z)$ maps the exterior of the unit circle onto the left half-plane. Therefore, to check the stability of the continuous filter, one can use the same criterion.

Example.

$$f(s) = 1 + 3s - s^2 + 2s^3.$$

In order to find out whether all the roots of this polynomial lie in the left half-plane (this is called the Hurwitz polynomial), we substitute $s = (1-z)/(1+z)$ and obtain

$$f^*(z) = (1+z)^3 + 3(1-z)(1+z)^2 - (1-z)^2(1+z) +$$

$$+ 2(1-z)^3 = -z^3 - 5z^2 + 13z + 5$$

5	13	-5	-1	× 5
-1	-5	13	5	× (-1)
24	60	-12	× 24	
-12	60	24	× (-12)	
432	2160			

$$|2160/432| > 1.$$

The polynomial $f^*(z)$ is not stable, therefore $f(s)$ is not the Hurwitz polynomial.

8. Rational Interpolation

Let $\{\beta_1, \beta_2, ..., \beta_s\}$ and $\{\lambda_1, ..., \lambda_s\}$ be sets of elements of an arbitrary field. It is required to find a rational function ϕ of the smallest degree, which assumes the present value at the points β_i:

$$\phi(\beta_i) = \lambda_i, \quad i = 1, ..., s, \tag{2.31}$$

$$\phi(z) - \lambda_i \equiv 0 \pmod{(z - \beta_i)}.$$

More generally, it may be assumed that the residues modulo $(z - \beta_i)^{m_i}$ are given:

$$\phi(z) = s_i(z) \pmod{(z - \beta_i)^{m_i}}, \quad i = 1, ..., s. \tag{2.32}$$

It is required to find a function ϕ of the smallest degree, which satisfies conditions (2.32).

By the Chinese remainder theorem one can find a polynomial $R(z)$ satisfying conditions (2.31)

$$g(z) = \prod_i (z - \beta_i),$$

$$g_i(z) = \frac{g(z)c_i}{(z - \beta_i)},$$

where c_i is such that $g_i(\beta_i) = 1$,

$$R(z) = \sum_i \lambda_i g_i(z).$$

It is now easy to find the function ϕ by the continued fraction algorithm.

$$\phi = \psi / f \equiv R(z)(\mathrm{mod}\,(g(z))),$$

$$\deg f \leqslant s/2,$$

$$\deg \psi < s/2.$$

In the general case (2.32), we consider the polynomials

$$g(z) = \prod_i (z - \beta_i)^{m_i}, \quad \deg g = n = \sum m_i,$$

$$g_i(z) = \frac{g(z)}{(z - \beta_i)^{m_i}}, \quad i = 1,\dots,s.$$

We choose polynomials $c_i(z)$ of degree $\deg c_i(z) < m_i$ such that

$$g_i(z)c_i(z) \equiv 1 \ (\mathrm{mod}\,(z - \beta_i)^{m_i}). \tag{2.33}$$

Then

$$R(z) = \sum c_i(z)g_i(z)s_i(z), \tag{2.34}$$

is a polynomial of degree smaller than n which satisfies condition (2.32). If $s_i(z)$ is expanded in powers of $(z - \beta_i)$

$$s_i(z) = a_0 + a_1(z - \beta_i) + \dots + a_{m_i - 1}(z - \beta_i)^{m_i - 1},$$

then coefficients a_j can be treated in the case of zero characteristic as the values of derivatives of $s_i(z)$ to within absolute terms.

Thus in this case the interpolation problem can be stated as follows. Given the values of a function and its derivatives

$$f(\beta_i),\ f'(\beta_i),...,f^{(m_i-1)}(\beta_i),\ i = 1,...,s,$$

find the function ϕ with the smallest number of poles, which satisfies those conditions (the Hermite interpolation). Thus, to solve this problem, we first find the polynomial $R(z)$ and next determine the function ϕ by the continued fraction algorithm.

9. Numerical Encoding

It is rather difficult to realize the error-correcting codes based on finite fields on a general-purpose computer since every algebraic operation in a finite field is implemented as a standard 'micro-routine'.

Here we shall consider a method of encoding numerical arrays in computer memory, which uses ordinary arithmetic of real (rational) numbers.

Let us consider a vector of length n over the rational field Q (a numerical array). An error of multiplicity no greater than t is said to occur if no greater than t numbers of the array (the vector components) have been distorted. Therefore, in this case the error corresponds to the distortion of several bits in a single machine word. We renumber the components of the vector by $1,2,...,n$ and put this vector into correspondence with the rational function

$$\zeta_x(z) = -\sum \frac{a_i}{z - \frac{1}{i}} = m_1 + m_2 z + ...$$

where

$$m_k = \sum_{i_i} a_i i^k,\ k = 1,2,....$$

The numbers m_k are called the moments of x. If the vector has t nonzero components, then the corresponding function is of degree t and the moments make up a linear recurrent sequence of order t.

To identify this sequence, it suffices to know the first $2t$ moments.

The code is the set of vectors such that $m_1 = m_2 = ... = m_{2t} = 0$. If x is a code vector and some (no greater than t) numbers are distorted, i.e., the error vector having no more than t nonzero components is added to a vector, then $2t$ moments of the sum $x + e$ coincide with the moments of the vector e and the error vector itself can be determined uniquely from these moments.

The error-checking matrix is

$$
\begin{bmatrix}
& 1 & 2 & ... & n & \\
& 1 & 2^2 & ... & n^2 & \\
\cdots\cdots\cdots\cdots\cdots\cdots\cdots\cdots\cdots\cdots\cdots \\
& 1 & 2^{2t} & ... & n^{2t} &
\end{bmatrix}
$$

i.e., consists of the points of a normal rational curve over Q.

If the code vector is identified with the polynomial $U(z) = a_1 z + ... + a_n z^n$, then the fact that the corresponding moments are equal to zero is equivalent to the conditions

$$U'(z)|_{z=1} = 0$$

$$U''(z)|_{z=1} =$$

$$\cdots\cdots\cdots\cdots\cdots$$

$$U^{2t}(z)|_{z=1} = 0.$$

This means that

$$U'(z) \equiv 0 (\mathrm{mod}\,(z-1)^{2t})$$

$$U(z) = (z-1)^d D(z) + C,$$

where $D(z)$ is a polynomial of degree less than $k = n - 2t$, $d = 2t + 1$. Since $U(0) = 0$, we obtain $C = D(0)$. Thus the code vector corresponds to the polynomial

$$U(z) = (z-1)^d D(z) + D(0).$$

Hence we obtain the following encoding method. The initial vector $(d_0, d_1, ..., d_{k-1})$ is identified with the polynomial

$$D = d_0 + d_1 z + ... + d_{k-1} z^{k-1}.$$

D is multiplied by $(z-1)^d$ and added to d_0. The multiplication by $(z-1)^d$ can readily be performed by evaluating the differences of order d.

Example. $t = 2$, the initial array is the set of 6 numbers

$$(11, -7. -2, 5, 8, 1).$$

We write it as a column and border with zeros. Further, the difference between neighbouring numbers are evaluated

0	0	0	0	0	0
11	-11	11	-11	11	-11
-7	18	-29	40	-51	62
-2	-5	23	-52	92	-143
5	-7	2	21	-73	165
8	-3	-4	6	15	-88
1	7	-10	6	0	15
0	1	6	-16	22	-22
	0	1	5	-21	43
		0	1	4	-25
			0	1	3
				0	1
					0

The encoding array is

$$(62, -143, 165, -88, 15, -22, 43, -25, 3, 1) \underset{\leftarrow}{\rightarrow} (62z - 143z^2 + ...).$$

Assume that after distortion has occurred, it takes the form

$$(62,1,3,-88,15,-22,43,-25,3,1).$$

We evaluate the moments

$$m_1 = -198;$$

$$m_2 = -882;$$

$$m_3 = -3222; \; m_4 = -10818.$$

We apply to the polynomial $R(z) = -18(11+49z+179z^2+601z^3)$ (the syndrome) the continued fraction algorithm and obtain the fraction

$$\phi(z) = \frac{18z - 33}{z^2 - z(5/6) + 1/6} = \frac{\psi(z)}{f(z)}.$$

In this case the roots of the polynomial f belong to the set $1, 1/2, 1/3, ..., 1/n$. The roots of the dual polynomial are the corresponding natural numbers:

$$f^*(x) = z^2 - 5z + 6 = (z-2)(z-3).$$

The values of these roots locate the error place. We now evaluate the values of errors: $a_2 = 144, a_3 = -162$.

We correct the errors and return to the initial numerical array, performing calculation (from bottom upward) according to the following scheme.

	11	-11	11	-11	11
62	-51	40	-29	18	-7
-143	92	-52	23	-5	-2
165	-73	21	2	-7	5
-88	15	6	-4	-3	8
15	0	6	-10	7	1
-22	22	-16	6	1	0
43	-21	5	1	0	
-25	4	1	0		
3	1	0			
1	0				
0					

The correction of a single error is particularly simple. In this case the ratio of moments $\mu = m_2 / m_1 = i$ is equal to the number of the distorted component and m_1 / μ determines the magnitude of the distortion.

Example 3. $t = 1$.

Assume that the distorted array is

$$(2, 20, -33, 14, 3, 20, -52, 56, -7, -33, 23, -4).$$

We evaluate the moments: $m_1 = 30$, $m_2 = 180$.

$$\mu = 180 / 30 = 6; \, a = 30 / 6 = 5.$$

It follows that 5 has to be subtracted from the 6th (from the left) number of the array.

The code in the letter example is a numerical counterpart of the Hamming code. In order to correct t or less numbers of the array, one has to add at least $2t$ redundant numbers to the initial array. Therefore, the described numerical codes are optimal as far as redundancy is concerned.

The encoded arrays can be summed (subtracted) and multiplied by a

0	0	0	0
4	-4	4	-4
10	-6	2	2
-2	12	-18	20
1	-3	15	-33
5	-4	1	14
7	-2	-2	3
-8	15	-17	15
12	-20	35	-52
11	1	-21	56
-4	15	-14	-7
0	-4	19	-33
	0	-4	23
		0	-4
			0

constant. Those operations are most frequently encountered in computers oriented at economic applications.

Comments

The *p*-adic analogs of the Diophantine approximations are studied in the Mahler's book [8]. The shift registers, linear recurrent sequences, and linear filters are developed in code theory owing to E. Prange who suggested the cyclic codes.

CHAPTER 3

Algebraic Curves

1. Affine and Projective curves

Let $F(X, Y)$ be a polynomial of two variables over the field F_q. A point (a,b) lying in the plane is called the root of the polynomial if $F(a,b) = 0$. All these roots can be found by enumeration and define a plane affine curve. Usually, one considers all points with coordinates in the algebraic closure of the field F_q, i.e., $a, b \in F_{q^m}$, $m = 1, 2, \cdots$. Points of the curve such that $(a,b) \in F_q$ are said to be rational over F_q. A projective curve is defined as a set of points lying in the projective plane and nullifying the form $F(X, Y, Z)$. Every such a form corresponds to three affine curves resulting from dehomogenization:

$$F(1, Y, Z), \ F(X, 1, Z), \ F(X, Y, 1).$$

Conversely, an affine curve converts to the projective one under homogenization: $Z^d F(X / Z, Y / Z)$, where d is the degree of F.

    Examples. The affine projective curve $F = Y^2 - X^2(X+1)$ is associated with the 'projective closure' $Y^2 Z - X^3 - X^2 Z$; the projective curve

$X^5 + Y^5 - Z^5$ is associated with the affine curve $X^5 + Y^5 = 1$.

A curve is called irreducible (or absolutely irreducible) if $F(X,Y)$ has no mul-
tipliers different from a constant in any extension F_{q^m} of the field F_q.

2. Intersection of Curves. Bezout Theorem

In order to find points of intersection of two curves F and G, it is needed to eliminate an unknown from the equations of these curves. Recall that the resultant
$R(\zeta, \eta)$ of two polynomials

$$\zeta = a_0 T^n + a_1 T^{n-1} + \cdots + a_n,$$

and

$$\eta = b_0 T^m + b_1 T^{m-1} + \cdots + b_m,$$

is the determinant

$$
\left|
\begin{array}{c}
\left.\begin{array}{ccccccc}
a_0 & a_1 & & \cdots & a_n & & \\
 & a_0 & a_1 & \cdots & & a_n & \\
 & & a_0 & a_1 & \cdots & a_n &
\end{array}\right\} m, \\
\left.\begin{array}{cccccccc}
b_0 & b_1 & \cdots & b_m & & & \\
 & b_0 & b_1 & \cdots & b_m & & \\
 & \cdots & \cdots & \cdots & \cdots & & \\
 & & b_0 & \cdots & b_1 & \cdots & b_m
\end{array}\right\} n.
\end{array}
\right.
$$

The basic theorem on the resultant reads: $R(\zeta, \eta) = 0$ if and only if
$a_0 = 0 = b_0$ or ζ and η has a common root. Now let $F(X,Y,Z)$ and
$G(X,Y,Z)$ be two plane curves without common components, i.e., F and G are
homogeneous forms of degrees m and n without common divisors. We
set $E(X,Z) = R_y(F,G)$, i.e., the forms are treated as polynomials of one varialbe
Y with coefficients of the field $K(X,Z)$. Obviously, the form $E(X,Z)$ is of
degree mn. On setting $Z = 1,0$, we obtain a polynomial $F(X)$ whose roots
determine the points of intersection of F and G. Since the roots can be multiple,

every common point P of the curves F and G is associated with a natural number $I(P,F \cap G)$ which is called the intersection number of the point P. The fact that the degree is equal to mn leads to the Bezout theorem

$$\sum_P I(P,F \cap G) = mn,$$

(two projective curves of degrees m and n without common components intersect at mn points, multiplicity being takes into account). The formal expression $\sum I(P)P$ is called the divisor of intersection of two curves.

If the point P is known, then the corresponding intersection number can be determined by a simple iterative algorithm (without evaluating the resultant). (This algorithm will be described in Section 3). In the general case the roots of the polynomial $f(X)$ are usually found by the Berlekamp method. The idea of this algorithm is as follows. We evaluate the polynomials $X^{q(i-1)}$ $(i=0,1,...,d-1)$ reduced modulo $f(X)$. The result is a $(d \times d)$ matrix $A = \|a_{ij}\|$, where

$$X^{qi} \equiv \sum_{j=0}^{d-1} a_{i-1,j-1} X^j \pmod{f(X)}.$$

Every vector of the zero space of matrix $A - I$, where I is a unit matrix, can be expressed as a polynomial satisfying the congruence

$$[g(X)]^q - g(X) \equiv 0 \pmod{f(X)},$$

and, since

$$g(X)^q - g(X) = \prod_{S \in F_q} [g(X) - S],$$

it follows that the greatest common divisor $(f(X),g(X)-S)$ divides $f(X)$. In this way $f(X)$ is factorized into irreducible multiplicators by the Euclidean algorithm.

3. Singular Points of a Curve

A point $P = (a,b)$ of a curve F is called singular, if all the derivatives $F_X(P)$, $F_Y(P)$, $F_Z(P)$ are zero at this point. Otherwise a point is called simple. If all points are simple, then the curve is called nonsingular (or smooth).

Example. For the curve $Y^2Z = X^3 - X^2Z$ over a field of characteristic 2, we have

$$F_X = X^2 = 0, \; F_Y \; 0, \; F_Z = (Y-X)^2 = 0,$$

hence $(0,0,1)$ is a singular point, while all the other points are simple.

Let $P = (0,0,1)$. We write $F(X, Y, 1) = F_m + F_{m+1} + \cdots F_n$, where n is the degree of the curve, $F_m \neq 0$, F_i is a form of two variables X and Y of degree i. The number m is called the multiplicity of F at P and we use the notation $m = m_p(F)$. If $P = (a,b, 1)$, then $m_P(F)$ is defined as $m_{(0,0,1)}(F^T)$, where $F^T = F(X+a, Y+b, 1)$.

Obviously, $P \in F$ is equivalent to $m_P(F) > 0$ and P is a simple point if $m_P(F) = 1$. Since F_m is a form of two variables, it follows that $F_m = \Pi L_i^{r_i}$, where L_i are different lines. They are called tangents to F at P, while r_i is called the multiplicity of a tangent. If all m tangents are different ($r_i = 1$), then the singular point is ordinary. In our example $F_m = (Y-X)^2$ hence the point $P = (0,0,1)$ of multiplicity 2 has the double tangent $Y - X$. If the same curve is considered over a field of a characteristic unequal to 2, then $F_X = 3X^2 = 0$, $F_Y = 2YZ = 0$, $F_Z = Y^2 - X^2 = 0$, the point P is a unique singular point as before, but in this case it is ordinary point with two different tangents $Y - X$ and $Y + X$.

Two curves F and G are said to be transversal at P if P is a simple point of both curves and the tangent to F at this point differs from the tangent to G at the same point. In this case the intersection number of this point is 1.

In the general case

$$I(P, F \cap G) \geq m_p(F)m_p(G),$$

and the equality takes place if and only if F and G have no common tangents at the point P.

Moreover, the intersection number possesses the following properties:

1) If $F = \Pi F_i^{r_i}$, $G = \Pi G_j^{s_j}$, then

$$I(P, F \cap G) = \sum r_i s_j I(P, F_i \cap G_j)$$

2)

$$I(P, F \cap G) = I(P, F \cap (G + AF))$$

Using these properties, assume that F and G are affine curves and $P = (0,0)$. If this is not the case, it is always possible to perform an affine transformation of the coordinates such that $T(Q) = P$. Then $I(Q, F^T \cap G^T) = I(P, F \cap G)$. Let $F(X, 0)$ and $G(X, 0)$ be polynomials of degrees r and s $(r \leqslant s)$.

Case 1. $r = 0$. Then Y divides F, hence $F = YH$ and

$$I(P, F \cap G) = I(P, Y \cap G) + I(P, H \cap G).$$

If $G(X, 0) = X^m(a_0 + a_1 X + ...)$, $a_0 \neq 0$, then

$$I(P, Y \cap G) = I(P, Y \cap G(X, 0)) = I(P, Y \cap X^m) = m.$$

Since $P \in G$, $m > 0$, it follows that the problem reduces to evaluating $I(P, H \cap G)$, where $\deg H < \deg F$.

Case 2. $r > 0$. We multiply F and G by constants such that $F(X, 0)$ and $G(X, 0)$ become unitary. Let $H = G - X^{s-r}F$. Then $I(P, F \cap G) = I(P, F \cap H)$ and $\deg H(X, 0) = t < s$. Repeating this process (and interchanging F and H when $t < r$), we come to Case 1 in a finite number of steps.

Example. We find the intersection number of the point $(0,1,0)$ where curve $F = ZY^2 + Z^2Y + X^3 + X^2Z$ intersects with the curve ZX. On setting $Y = 1$, we obtain

$$F = Z^2 + Z(X^2 + 1) + X^3, \ G = ZX;$$

$$F(X, 0) = X^3, \ G(X, 0) = 0; \ r = 0; \ s = 3;$$

$$I(P, F \cap G) = I(P, F \cap Z) + I(P, F \cap X);$$

$$I(P, Z \cap F) = I(P, Z \cap F(X, 0)) = I(P, Z \cap X^3) = 3.$$

It remains for us to evaluate $I(P,F \cap H)$, where $H = X$. Since $H(X,0) = X$, it follows that

$$I(P,F \cap X) = I(P,Z(Z+1) \cap X) = I(P,Z \cap X) +$$

$$+ I(P,(Z+1) \cap X) = 1,$$

in view of $I(P,(Z+1) \cap X) = 0$.

Thus the intersection number is 4. The straight line $Z = 0$ is tangent to the curve F at P, this tangent is triple: $I(P,F \cap Z) = 3$ (in this case P is said to be a point of inflection). The line $X = 0$ intersects F at three different points: $P = (0,1,0)$, $Q_1 = (0,0,1)$, and $Q_2 = (0,1,1)$. Therefore, the total intersection number is equal to 4.

Every singular point is a point of intersection of four curves. To find those points one may evaluate the resultants

$$R_Y(F,F_X) = E_1(X,Z),$$

and

$$R_Y(F_Y,F_Z) = E_2(X,Z),$$

and then apply the Euclidean algorithm to find the greatest common denominator (E_1,E_2) for $Z = 1,0$.

4. Linear Systems of Curves

Let $M_1,...,M_N$ be all possible monomials of X,Y,Z of degree d, i.e., X^d, $X^{d-1}Y$, $X^{d-1}Z$, $\cdots$. Obviously, $N = (d+1)(d+2)/2$. Specifying a curve of degree d is the same as choosing $a_1,a_2,...,a_N$ not all equal to zero and setting $F = \sum a_i M_i$ to within the multiplication by a constant. In other words, all the curves of degree d make up a projective space P^{N-1} of dimension $d(d+3)/2$. Specifically, all the straight lines make up P^2, concis (the curves of degree 2) P^5, the cubics (the curves of degree 3) P^9, quartics (the 4th degree) P^{14}, the quintics (the 5th degree) P^{20}. Moreover, if two curves differing by a constant factor are regarded as different and if a zero curve is added, then all the curves of degree d make up an affine space A^N, where $N = (d+1)(d+2)/2$.

We fix a point Q. The all the curves of a given degree passing through this point make up a hyperplane. The change of coordinates X, Y, Z (a projective transformation) leads to a projective transformation of coordinates in the linear manifold of curves $P^{d(d+3)/2}$.

Let the point Q be $(0,0,1)$. If curve $F = \sum F_i(X,Y)Z^{d-i}$ passes through this point with multiplicity $m_Q(F) \geq r$, it means that all the forms $F_i(X,Y)$ for $i < r$ are zero, i.e., all the coefficients of monomials $X^i Y^j$ with $i+j < r$ are equal to 0. All in all there are $r(r+1)/2$ such coefficients. Therefore, all the curves of degree d such that $m_Q(F) \geq r$ make up a linear manifold of dimension $d(d+3)/2 - r(r+1)/2$. In this case the point Q of multiplicity r imposes $r(r+1)/2$ linear conditions. The point Q may be arbitrary since it can always be converted to the point $(0,0,1)$ by a projective transformation.

Now let a number of points $P_1,...,P_n \in P^2$ be given, which are associated with natural numbers $r_1,...,r_n$ (i.e., a positive divisor $D = \sum r_i P_i$ is specified). We denote by $V(d,D)$ the linear system of curves of degree d passing trhough D, i.e., curves such that $m_{P_i}(F) \geq r_i$, $i = 1,...,n$.

It is clear that

$$\dim V(d,D) \geq d(d+3)/2 - \sum r_i(r_i+1)/2$$

where the equality holds when all the points P_i impose independent conditions. The latter is the case if d is sufficiently large.

Theorem 4. If $d \geq (\sum r_i) - 1$, then

$$\dim V(d,D) = d(d+3)/2 - \sum r_i(r_i+1)/2.$$

Idea of the proof: if all $r_i = 1$, then it is always possible to find a straight line L_i passing through P_j for $i \neq j$ since we consider the algebraically closed field of constants (which is consequentely infinite). Moreover, one can find a line L_0 which does not pass through any point P_i. Then the form $L_1 L_2 \cdots L_{n-1} L_0^{d-n+1}$ defines a curve of degee $d \geq n-1$ passing through $n-1$ points and not passing through the remaining point. This exactly means that the conditions imposed by points P_i are independent. In the general case when $r_i > 1$ the proof is conducted by induction.

5. Fundamental Noether Theorem

In the previous section we discussed how one can find all the curves of a given degree passing through a positive divisor D. Now let D be the divisor of intersection of two curves F and G of degrees m and n such that $D = \sum I(P, F \cap G)P$. It truns out that under some rather general conditions there is no necessity to determine the intersection points and the corresponding numbers: one can straightforwardly write that all the curves of degree d which pass through D are of the form $AF + BG$ where A and B are forms of degrees $d - m$ and $d - n$.

The idea underlying the proof of this theorem can easily be explained in the simplest case when all mn points of intersection are different.

First, let $d \geqslant mn - 1$. Then according to Theorem 4 the dimension of all curves of degree d in the space A^N, which pass through mn points, is exactly $(d+1)(d+2)/2 - mn$. We compare this number with the dimension of all forms $AF + BG$ of degree d. Specifying such a curve is equivalent to choosing a point in A^{N_1} corresponding to the form A and a point in A^{N_2} corresponding to the form B, where

$$N_1 = (d - m + 1)(d - m + 2)/2, \quad N_2 = (d - n + 1)(d - n + 2)/2.$$

In other words, a linear mapping from $A^{N_1} \times A^{N_2}$ to the space of all forms $AF + BG$ is defined. The kernel of such a mapping is not zero, it includes the space of points $(XG, -XF)$, where X is any form of degree $d - m - n$. It is clear that all these points make up a space of dimension

$$s = (d - m - n + 1)(d - m - n + 2)/2.$$

Thus the dimension of the space of all forms $AF + BG$ is no less than

$$N_1 + N_2 - s = (d + 1)(d + 2)/2 - mn,$$

which proves the theorem for $d \geqslant mn - 1$.

We now reduce the degree d: if the theorem holds true for curves of degree d, then it is also true for curves of degree $d - 1$. Let S be of degree $d - 1$ and pass through $F \cap G$, and let L be an arbitrary straight line not passing through $F \cap G$. Then LS is a curve of degree d satisfying the induction hypothesis, therefore

$$LS = AF + BG = (A + XG)F + (B - XF)G.$$

The line L intersects F at m points which do not lie on G. Therefore, these points lie on B and consequently on $B - XF$. Similarly, L intersects G at n points which lie on $A + XG$. The curve X depends on $s \geqslant d - m - n + 1$ parameters and in the case $s \geqslant m + n$ it is always possible to choose X such that $A + XG$ passes through $d - m - n + 1$ more points of the line L different from the points of $L \cap F$ and $L \cap G$. Alltogether, $A + XG$ contains $d - m + 1$ points of the line L. But since the degree of this curve is $d - m$, it follows from the Bezout theorem that $A + XG$ contains the whole line L:

$$A + XG = LA', \quad \deg A' = d - m - 1.$$

But then $B - XF$ also contains the whole L:

$$B - XF = LB', \quad \deg B' = d - n - 1.$$

Hence

$$S = A'F + B'G.$$

Thus the theorem is true whenever $d \geqslant m + n$.

Finally, let $d = m + n - k$, $k > 0$, $d \geqslant m, n$. We choose a curve S' of degree k which does not pass through the points of intersection of $F \cap G$. Let S be the initial curve of degree d. Then SS' is a curve of degree $m + n$ passing through $F \cap G$. Then

$$SS' = AF + BG = (A + cG) + (B - cF)G,$$

where c is an arbitrary constant.

The curve S' intersects F at km points which lie on $B - cF$. If we choose c such that $B - cF$ intesects S' at one more point the $B - cF$ contains the entire

S'. It follows that

$$S = A'F + B'G.$$

We have proved the Noether theorem in the case when all mn points of intersection are different. In fact this theorem remains valid under more general conditions which can be derived in a similar manner but require that the 'freedom in the choice of parameters' should be taken into account in a more sophisticated way.

The Noether theorem is valid for a form H if at every point P of intersection of F and G one of the following conditions is satisfied:

1) F and G are transversal to one another at P and $P \in H$;

2) P is a simple point on F and

$$I(P, H \cap F) \geqslant I(P, G \cap F);$$

3) F and G have different tangents at P and

$$m_P(H) \geqslant m_P(G) + m_P(F) - 1.$$

Obviously, in this case we have

$$I(P, H \cap F) \geqslant m_p(H)m_p(F) \geqslant I(P, G \cap F) + m_p(F)(m_p(F) - 1).$$

4) P is an ordinary singular point on F of multiplicity r_p and

$$I(P, H \cap F) \geqslant I(P, G \cap F) + r_P(r_P - 1).$$

Example. It is required to find a curve passing through all q^2 points of the affine plane A^2 over the field F_q. Consider curves $F = X^q - XZ^{q-1}$ and $G = Y^q - YZ^{q-1}$. The curves intersect at q^2 points of A^2 therefore, the simplest Noether condition holds. Any curve passing through those points is of the form

$$H = A(X, Y, Z)(X^q - XZ^{q-1}) + B(X, Y, Z)(Y^q - YZ^{q-1}).$$

If the forms A and B are of degree 0, i.e., are constant, then H decomposes since in this case

$$H = aX^q + bY^q - (aX + bY) = (aX + bY)^q - (aX + bY),$$

because $a,b \in F_q$. Now let $A = X$, $B = Z$. Then

$$H = Y^q Z - YZ^q + X^{q-1} - X^2 Z^{q-1}.$$

On setting $Z = 1$, we obtain the affine curve

$$H = Y^q - Y + X^{q-1} - X^2,$$

which is absolutely irreducible. In order to find singular points, we write the set of equations

$$H_X = X^q - 2XZ^{q-1} = 0$$

$$H_Y = -Z_q = 0$$

$$H_Z = Y^q + X^2 Z^{q-2} = 0$$

$$H = 0.$$

We have a unique solution (0,0,0) which is not a point of the projective plane. Therefore, the curve H is nonsingular.

Thus the minimum degree of the irreducible smooth curve passing through all the points of the projective plane over F_q is $q + 1$.

6. Rational Functions and Divisors

Let F be an irreducible curve. The ratio of two polynomials $f = a(X,Y)/b(X,Y)$ is called the rational function on the affine curve, note that two ratios f and g are identified if $f - g \equiv 0 \pmod{F(X,Y)}$. In the case of a projective curve, the function is the ratio of two forms of the same degree $f = A(X,Y,Z)/B(X,Y,Z)$ to within factorization modulo $F(X,Y,Z)$. A function is defined at point P, if there exists its representation $f = A/B$, where $B(P) \neq 0$. In this case one can evaluate the function at a point: $f(P) = A(P)/B(P)$.

Example. Consider the function $f = (Y^2 + YZ)/ZX$ on the curve $F = Y^2 Z - YZ^2 + X^3 - X^2 Z$ over the field F_2. Evaluate the function at the point $P = (0,0,1)$. We note that

$$\frac{Y^2 + YZ}{ZX} - \frac{X(X - Z)}{Z^2} \equiv 0 (\text{mod } F),$$

therefore f admits the representation

$$f = X(X - Z)/Z^2,$$

and its value at P is 0.

For the function $f = A / B$ we find the divisors D_A and D_B of intersection of the form A and B with the curve F. These divisors can have a common component:

$$D_A = H + G_0,$$

$$D_B = H + G_\infty.$$

Here G_0 and G_∞ have no common points. The divisor $G_0 = \sum m_p P$ is called the divisor of zeros and $G_\infty = \sum n_Q Q$ the divisor of poles. Accordingly, the point P is called a zero of f of multiplicity m_p, and the point Q is called a pole of multiplicity n_Q.

The formal sum $D = \sum m_p P$, where m_p is an integer (possibly negative) and $m_p = 0$ for all points except for a finite number of them, is called the divisor on the curve F.

The value deg $(D) = \sum m_p$ is called the degree of the divisor. Two divisors can be added (subtracted) by performing the corresponding operation over the numbers m_p at the same point P. For example, if $D_1 = P_1 + 2P_2 - P_3$, $D_2 = P_1 - 3P_3 + P_4$, then $D_1 - D_2 = 2P_2 + 2P_3 - P_4$. Finally, we introduce the order relation

$$\sum n_i P_i = D \geqslant G = \sum m_i P \text{ it means } n_i \geqslant m_i \text{ for all } i,$$

for the divisors.

If all $m_p \geqslant 0$, then the divisor is positive (effective). Those are the divisors which have been considered so far as the divisors of intersection of two curves.

Thus every function can be associated with a divisor

$$(f) = D_A - D_B = G_0 - G_\infty = \sum n_p P.$$

The value n_P is called the order of a function at P and the notation $\text{ord}_P(f)$ is used. Since D_A and D_B are of the same degree (by the Bezout theorem), it follows that deg $(f) = 0$: every function has the same number of zeros and poles, multiplicity being taken into account.

A function is uniquely defined by its divisor to within a constant multiplier: if $(f) = (g)$, then $(f/g) = (f) - (g) \equiv 0$ i.e., the function f/g has neither zeros, nor poles and consequently is a constant.

We have defined zeros and poles of a function on the basis of the specific representation $f = A/B$. Now let another represenatation $f = A_1/B_1$ be given such that $A/B - A_1/B_1 \equiv 0(\text{mod } F)$, $B_1 A = A_1 B + XF$ for a form X. Then

$$I(P, F \cap (B_1 A)) = I(P, F \cap (A_1 B))$$

$$I(P, F \cap B_1) + I(P, F \cap A) = I(P, F \cap A_1) + I(P, F \cap B)$$

in view of properties (1) and (2) of the intersection number.

Therefore,

$$I(P, F \cap A) - I(P, F \cap B) = I(P, F \cap A_1) - I(P, F \cap B_1).$$

This means that the divisor of a function is actually independent of its specific representation as a ratio of two forms.

Two divisors D and D' are called linearly equivalent if $D' - D = (f)$ for a function f. In this case we write $D' \equiv D$. This relation is in fact an equivalence relation, i.e., it is symmetric, reflexive, and transitive. Moreover, it possesses the following properties:

1) $D \equiv 0$ if and only $D = (f)$;

2) If $D \equiv D'$, then $\deg(D) = \deg(D')$;

3) If $D \equiv D'$ and $D_1 \equiv D_1'$, then

$$D + D_1 \equiv D' + D_1'$$

4) $D \equiv D'$ implies that there exist two curves G and G' of the same degree such that

$$D + \operatorname{div}(G) = D' + \operatorname{div}(G').$$

7. Riemann-Roch Problem

The Riemann-Roch problem admits 3 equivalent formulations:

1) Given a divisor D, one has to find all effective divisors D' which are linearly equivalent to $D: D' \equiv D$. In the case of the finite field of constants it is required to find the number of the divisors making up a single class, while in the case of an infinite field it is required to find the number of parameters on which this class depends. Moreover, an algorithm is needed to find all these divisors.

2) Let $D' \equiv D$, hence $D' - D = (f)$, $D' = D + (f) \geqslant 0$. If $D = \sum n_P P$, this means that $(f) \geqslant -D$, therefore

$$\operatorname{ord}_P(f) \geqslant -n_P. \tag{3.1}$$

The set of all functions satisfying condition (3.1) at every point P is called the space associated with the divisor D and denoted by $L(D)$. In the case when D is an effective divisor, $L(D)$ consists of the functions such that all poles lie in D and the multiplicity of each pole is no greater than n_P. If D consists of positive and negative components, i.e.,

$$D = D^- + D^- = D^+ + (-\sum m_Q Q),$$

then at Q a function of $L(D)$ must have zero with multiplicity no less than m_Q. Obviously, $L(D)$ is a vector space over the field of constants k. The dimension of this space is denoted by $l(D)$, The Riemann-Roch theorem deals with evaluation of $l(D)$.

3) Consider the linear system of curves

$$\lambda_0 f_0(X, Y, Z) + \lambda_1 f_1(X, Y, Z) + \cdots + \lambda_{r-1} f_{r-1}(X, Y, Z) = 0,$$

where $f_0, \ldots, f_{r-1}$ are linearly independent forms of the same degree and $\lambda_0, \ldots, \lambda_{r-1} \in k$. All the curves of the linear systems cut out effective divisors at the initial curve. The set of such divisors with $\lambda_0, \ldots, \lambda_{r-1}$ running the field k is called the linear series. The curves $f_0, \ldots, f_{r-1}$ and F can pass through a common set of points (a divisor), in which case all the divisors of the linear series have a common component which can be omitted. The remaining divisors are different. Indeed, if two divisors of the linear series coincided, then the corresponding curves would differ by a constant multiplier, which contradicts the assumption that the basic functions $f_0,...,f_{r-1}$ are linearly independent. Thus all the divisors of the linear series are in a one-to-one correspondence with the space P^{r-1}. The notation g_n^r is adopted for the linear series where n is the degree of the divisor.

On dividing all the forms f_i of the linear system by one of the forms, say f_0, we obtain the linear system of functions

$$\lambda_0 + \lambda_1 \phi_1 + \cdots + \lambda_{r-1} \phi_{r-1} = 0$$

If D is the divisor of the form f_0, then all the functions of the system belong to the space $L(D)$. In the specific case, when the system coincides with the space $L(D)$, it is called complete. In this case $\{1, \phi_1, ..., \phi_{r-1}\}$ is a basis of the space $L(D)$ and the linear series coincides with the linear equivalence class of the divisor D.

Thus an arbitrary linear series is a subclass of equivalent divisors and can be uniquely extended to a complete linear series. It is also clear that if two complete linear series contain a common divisor, then the series coincide.

Thus, the notion of the space associated with a divisor is equivalent to the notion of the complete linear system. But such a linear system can readily be built up by using the Noether theorem.

Let the initial curve F of degree s have ordinary singular points Q of multiplicity r_Q.

If two divisors D and D' are linearly equivalent, then there exist two forms H and H' of the same degree such that

$$D + \text{div}(H) = D' + \text{div}(H').$$

Let G be an arbitrary curve of degree l which passes through every singular point Q of the curve F with multiplicity no less than $r_Q - 1$ (such a curve is called adjoint to F).

We introduce the divisor $E = \sum(r_Q - 1)r_Q Q$. Then

$$\text{div}(GH) = \text{div}(G) + \text{div}(H) = D' + \text{div}(H') - D + \text{div}(G).$$

If G also passes through D, i.e., div $(G) = D + E + A$, then div $(GH) \geqslant \text{div}(H') + E$, since at every singular point we have

$$I(Q, G \cap F) \geqslant m_Q(G)m_Q(F) = (r_Q - 1)r_Q$$

We now apply the Noether theorem to the curves $F \cap H'$ and GH. At every point of interesection of the curves H' and F the condition (4) of Section 5 holds, therefore the curve GH can be represented in the form

$$GH = FF' + G'H'$$

for some forms F' and G'. It follows that

$$\text{div}(G') = \text{div}(GH) - \text{div}(H') = D' + E + A.$$

Thus G' is an adjoint curve of the same degree as G (since it passes through the divisor E).

We have thus proved the Noether theorem on the residue divisor.

Theorem 5. Let $D \equiv D'$ and let G be an adjoint curve of degree l such that div $(G) = D + E + A$, where A is a 'residue' divisor. Then there exists an adjoint curve of degree l such that

$$\text{div}(G') = D' + E + A.$$

The theorem leads to the following algorithm of constructing a basis of the space $L(D)$ in the case when F is an irreducible curve with ordinary singular points.

1) Find a basis of the adjoint curves of degree l. To this end one should singular out in the linear system of curves of degree l (which has the dimensionality $l(l+3)/2$) the curves G passing through every singular point Q of the curve F with multiplicity less by one: $m_Q(G) \geqslant r_Q - 1$. Every such a point imposes $(r_Q - 1)r_Q / 2$ conditions and therefore the desired system is of the dimension

$$r \geqslant l(l+3)/2 - \sum(r_Q - 1)r_Q / 2.$$

This is true in the cases when $l < m$, where m is the degree of the curve F.

If $l \geqslant m$, then there exist curves of degree l which contain the entire F. All the curves make up a linear space of dimension

$$((l-m)(l-m+3)/2) + 1$$

(for example, for $l = m$ there exists a 1-dimensional space of the curves cF). Therefore, one should first perform factorization by this space. The resulting system has dimension

$$r \geqslant l(l+3)/2 - \sum(r_Q-1)r_Q/2 - (l-m)(l-m+3)/2 - 1 (l \geqslant m).$$

2) Find the curves of this system passing through the given divisor $D = \sum n_P P$. This corresponds to the introduction of new linear conditions whose number do not exceed $\sum n_P(n_P+1)/2$.

Note that in the case of a nonsingular curve F, all the curves are adjoint, which simplifies the calculations.

3) Find an arbitrary adjoint curve G passing through the divisor D.

Let

$$\mathrm{div}\,(G) = D + E + A,$$

where A is a residue divisor.

We can always assume that A is of the form $A = \sum R_i$ where all the points R_i are different. Since $\deg(E) = \sum(r_Q-1)r_Q$, it follows that the degree of A is equal to

$$\deg(A) = ml - \sum(r_Q-1)r_Q - \deg(D).$$

4) Find all the adjoint curves of degree l passing through the residue divisor A. To this end one should solve the set of linear equations connected with the conditions imposed by this divisor. As a result we obtain a complete linear series g_n^r corresponding to the space $L(D)$. Here

$$r \geqslant l(l+3)/2 - \sum(r_Q-1)r_Q/2 - \deg(A) \text{ with } l < m \tag{3.2}$$

$$r \geqslant l(l+3)/2 - \sum(r_Q-1)r_Q/2 - \deg(A) - \tag{3.3}$$

$$- (l-m)(l-m+3)/2 - 1 \text{ with } l \geqslant m$$

$$n = \deg(D) = ml - \sum(r_Q - 1)r_Q - \deg(A). \tag{3.4}$$

Since all the curves of the system pass through the divisor A, this divisor is a common component of all the divisors in the linear series and it can be omitted. The linear series g_n^r contain D and all equivalent divisors.

Example. Consider the curve $F = Y^2Z - YZ^2 + X^3 - X^2Z$ over the field F_2 and the divisor $D = 4P$ where $P = (0,1,0)$. This is a nonsingular curve of degree $m = 3$. We construct the space $L(D)$ using the curves of degree $l = 2$. The straight line $Z = 0$ is a tangent to F and this is a triple tangent, hence $I(P, Z \cap F) = 3$. The line $X = 0$ intersects the curve F at 3 different points

$$P = (0,1,0), \ R_1 = (0,0,1), \ R_2 = (0,1,1).$$

Therefore, the conic $G = ZX$ intersects F over the divisor $D + R_1 + R_2$. Here $A = R_1 + R_2$ is a residue divisor. The linear system of conics passing through the poins R_1 and R_2 has the basis

$$\{X^2, Y^2 + YZ, XY, XZ\}.$$

On dividing all the basic curves by XZ, we obtain the functions

$$\{1, \ X/Z, \ (Y^2 + YZ)/ZX, \ Y/Z\}.$$

Thus in this case $l(D) = 4$ and the linear series g_n^r has the parameters $n = 4, r = 3$ (the notation g_n^r contains the projective dimension, which is smaller than the affine dimension by 1).

8. Riemann-Roch Theorem

We now clarify how r and n interrelate in the complete linear system g_n^r. To this end we eliminate the parameter l in the inequalitiees (3.2), (3.3), and (3.4). We introduce the value

$$g = (m-1)(m-2)/2 - \sum(r_Q - 1)r_Q/2,$$

which is called the genus of a curve.

Then (3.4) takes the form

$$n = 2g - 2 + m(l - m + 3) - \deg(A) \qquad (3.5)$$

and expression (3.3) converts to the inequality

$$r \geqslant g - 2 + m(l - m + 3) - \deg(A). \qquad (3.6)$$

Comparing (3.5) and (3.6) we obtain the inequality

$$r \geqslant n - g$$

for the case $l \geqslant m$. Note that the right-hand sides in (3.2) and (3.3) coincide for $l = m - 1$ and $l = m - 2$.

It remains for us to study the case $l \leqslant m - 3$.

Let $l = m - 3 - \alpha(\alpha \geqslant 0)$. Then it follows from (3.2) that

$$r \geqslant g - 1 - m\alpha + \alpha(\alpha + 3)/2 - \deg(A).$$

On substituting (3.5), we obtain

$$r \geqslant n - g + 1 + \alpha(\alpha + 3)/2,$$

hence in this case the inequality

$$r \geqslant n - g$$

holds also.

Thus the Riemann relation

$$l(D) \geqslant \deg(D) - g + 1$$

holds true.

In order to obtain an exact expression for $l(D)$, we consider separately the linear series cut eut by the adjoint curves of degree $(m - 3)$ (ϕ-curves). These series are called special and the number of linearly independent ϕ-curves passing through D is called the specialization index of dividor D denoted by $i(D)$. If $i(D) = 0$, then the divisor is called nonspecial, otherwise the divisor is special. In these terms the exact formula for $l(D)$ takes the form

$$l(D) = \deg(D) - g + 1 + i(D)$$

(the Riemann-Roch theorem).

The proof is conducted by induction on the degree of D.

The linear series cut by all curves of the $(m-3)$th degree is canonical and consists of canonical divisors. Each of the divisors possesses the following properties

$$\deg(w) = 2g - 2, \ l(w) = g.$$

The specialization index enjoys the following properties:

(i) $0 \leqslant i(D) \leqslant g$;

(ii) $i(D) = 0$ whenever $\deg(D) \geqslant 2g - 1$;

(iii) $i(D) = 0$ whenever $l(D) > \frac{1}{2}\deg(D) + 1$

(the Clifford theorem).

To compute $i(D)$, one has to write the linear system of ϕ-curves passing through D and to find the rank of this system.

Example. 1) For a rational curve $(g=0)$ we have $i(D) = 0$ for any divisor. Specifically, if $D = P$, then $\deg(D) = 1$ and $l(D) = 2$. Thus there is always a function with 1 pole at P.

2) For an elliptic curve $(g=1)$ we have $2g - 1 = 1$, therefore all the divisors are also nonspecial. If $D = P$, then $l(D) = 1$ and there is no function with a single pole. But if $D = P_1 + P_2$ or if $D = 2P$, then $l(D) = 2$. Thus there always exists a function with any two poles.

3) Consider a nonsingular quartic $(g=3)$. In this case, every ϕ-curve is a straight line, i.e., the canonical class coincides with P^2. The canonical divisor is the intersection of the quartic with a straight line and consists of $2g - 2 = 4$ points. If $D = P$, then $i(D) = 2$ is the dimension of the space of all curves passing through P. Therefore $l(D) = 1$. Consequently, there is no function with a single pole.

Now let $D = P_1 + P_2$. Since there exists a one-dimensional space of straight lines passing through two points, $l(D) = 1$ and there is no function with two poles.

Finally, $D = P_1 + P_2 + P_3$. If those three points do not lie on the same line, then the divisor D is nonspecial and $l(D) = 1$. In this case there is no function with poles P_1, P_2, P_3. But if all the three points are collinear, then $i(D) = 1$ and $l(D) = 2$.

In this case there exists a function with the three poles.

9. Quadratic Transformations

In the previous sections we discussed a curve with ordinary singular points and defined the value

$$g = (m-1)(m-2)/2 - \sum r_Q(r_Q-1)/2,$$

which is called the genus of the curve. This value plays a crucial role in the Riemann-Roch theorem.

The rational mapping of a curve F into a curve G is a pair of rational functions $\phi(X,Y)$ and $\psi(X,Y)$ such that, on substituting $X' = \phi$ and $Y' = \psi$, the polynomial $G(X',Y')$ is zero (for projective curves one needs forms $\phi_1(X,Y,Z)$, ϕ_2, ϕ_3 of the same degree rather than the rational functions ϕ and ψ. If there exists mutually inverse rational mappings of F and G into F, then the curves F and G are called birationally equivalent.

Under such a mapping linearly equivalent divisors convert to linearly equivalent ones, hence a complete linear series $L(D)$ corresponds to a complete linear series $L(D')$. A canonical series converts to a canonical series, so the genus of a curve does not change under a birational mapping. The converse generally speaking, is not true-two curves can have the same genus not being birationally equivalent, hence the genus is not a complete invariant. The simplest type of birational transformations is the projective equivalence, i.e., the change of coordinates. In this case forms ϕ_i are linear and generated linear nonsingular mapping of P^2 into itself.

Another important type of birational transformation is the standard quadratic Cremona transformation

$$\phi_1 = YZ, \quad \phi_2 = XZ, \quad \phi_3 = XY.$$

Combining projective and quadratic transformations, one can convert any plane curve to a plane curve with ordinary singularities. Specifically, this makes it possible to calculate the genus of any plane curve.

Let $P = (0,0,1)$, $P' = (0,1,0)$, $P'' = (1,0,0)$.

These are the basic points. The lines $L: Z = 0$, $L': Y = 0$, and $L'': X = 0$ are called exceptional. We denote by U the place P^2 without these lines.

The mapping

$$Q: P^2 - \{P,P',P''\} \to P^2$$

defined as

$$Q((X,Y,Z) = (YZ,XZ,XY)$$

is invertible on the set U:

$$Q(Q(X,Y,Z)) = (XZXY,YZXY,YZXZ) = (X,Y,Z)$$

and therefore, is a birational mapping onto itself.

If F is a curve in P^2, then Q converts it to the curve

$$F^Q = F(YZ,XZ,XY).$$

If F is of degree n, then F^Q is a curve of degree $2n$.

Let

$$m_P(F) = r, \ m_{P'}(F) = r', \ m_{P''}(F) = r''.$$

Then

$$F^Q = Z^r Y^{r'} X^{r''} F',$$

where F' is an irreducible curve of degree

$$\deg(F') = 2n - r - r' - r''$$

It is easily seen that

$$(F')' = F, \ m_P(F') = n - r' - r'', \ m_{P'}(F') = n - r - r'',$$

$$m_{P''}(F') = n - r - r'.$$

The curve F is said to be in a standard position if
1) any exceptional line is not a tangent to F at a basic point;
2) L is transversal to F at n different points which are not basic;

3) L' and L'' are transversal to F each at $n - r$ different nonbasic points.

If all these conditions are satisfied for F, then the corresponding curve F' has the following multiple points:

a) the multiple points of F lying in U convert to points of the same multiplicity of $F' \cap U$ an ordinary point moving to an ordinary one;

b) P, P', and P'' become ordinary singular points of F' with $m_p = n$, $m_{p'} = n - r$, $m_{p''} = n - r$ (in the standard position F does not pass through P' and P'', therefore, $m_{p'}(F) = m_{p''}(F) = 0$);

c) on $F' \cap I'$ or $F' \cap L''$ there can be only basic points; on $F' \cap L$ there can be also nonbasic points $P_1, ..., P_s$; then

$$m_{P_i}(F') \leqslant I(P_i, F' \cap L), \quad \sum I(P_i, F' \cap L) = r.$$

The value

$$g^*(F) = (n-1)(n-2)/2 - \sum r_Q(r_Q - 1)/2$$

is called the virtual genus of a curve. If all the points are ordinary, then g^* coincides with g. For an arbitrary curve we have

$$g^*(F) \geqslant g(F).$$

If F is in the standard position, then all its singular points lie in $U \cup \{P\}$ and $m_P(F) = r$. Therefore,

$$g^*(F) = (n-1)(n-2)/2 - r(r-1)/2 - \sum r_Q(r_Q - 1)/2,$$

where Q is not basic.

Passage to F' preserves the latter sum and, since

$$\deg(F') = 2n - r,$$

it follows that

$$g^*(F') = (2n - r - 1)(2n - r - 2)/2 - n(n-1)/2 -$$

$$(n-r)(n-r-1)/2 - 2 - \sum r_Q(r_Q - 1)/2 - \sum_{i=1}^{s} r_i(r_i - 1)/2,$$

where $r_i = m_{P_i}(Fprime)$ for nonbasic points P_i lying in $F' \cap L$. It follows that

$$g^*(F') = g^*(F) - \sum_{i=1}^{s} r_i(r_i - 1)/2.$$

Thus the quadratic transformation converts the point P in the standard position to an ordinary one. If new multiple points P_i appear, then the virtual genus decreases:

$$g^*(F') < g^*(F).$$

Let Q be a singular point of a curve, which is not ordinary. Then it is always possible to perform a projective transformation T converting the curve to the standard position and the point Q to $P = (0, 0, 1)$. We first shift the point Q to P. The curve takes the form

$$F = F_r(X, Y)Z^{n-r} + \cdots + F_n(X, Y)$$

where F_i is a form of degree i, $F_r \neq 0$.

We find two lines passing through P. The equation of such a line is

$$X = \lambda Y$$

hence every point of this line different from P is of the form $(\lambda, 1, t)$. In order that the polynomial

$$F_\lambda(t) = F_r(\lambda, 1)t^{n-r} + \cdots + F_n(\lambda, 1)$$

have $n - r$ different roots, it suffices that

$$F_r(\lambda, 1) \neq 0, \quad F_\lambda' = dF_\lambda(t)/dt$$

has no common roots with $F_\lambda(t)$. The discriminant of $F_\lambda(t)$ (the resultant of $F_\lambda(t)$ and $F_\lambda'\lambda(t)$), which is a polynomial of λ, has only a finite number of zeros. The remaining values of λ lead to the lines required by the standard position.

In the case of characteristic $p \neq 0$ it may turn out that the derivative of F_r identically equals zero, hence $F_\lambda'(t) = 0$. In this case

$$F = F_r(X, Y)Z^{p i_1} + \cdots + F_n(X, Y)$$

and therefore

$$n - r \equiv 0 (\bmod p).$$

We choose a point S of multiplicity $m = 0$ or 1 and convert the curve to the standard position using a projective transformation such that $T(0,0,1) = S$ and S does not lie on any basic line. We now perform the quadratic transformation and obtain a new curve of degree $n' = 2n - m$.

So we have

$$n' - r = 2n - m - r = n - r + n - m.$$

Therefore,

$$n' - r \equiv n - m (\bmod p).$$

One of the values $m = 0$ or $m = 1$ guarantees that $n' - r$ is not divided by p.

Once the lines L' and L'' passing through $n - r$ different points are chosen, one can choose a line L passing through n different points of the curve. There always exists a projective transformation converting L to Z, L' to Y, and L'' to X. The curve is thereby transformed to the standard position and one can perform the quadratic transformation centred at P. On repeating this procedure for each of N singular points, which are not ordinary, we obtain a curve with ordinary singularities in $N + g^*(F)$ steps.

10. Projective Model of the Linear System

Consider the linear system g_n^r

$$\lambda_0 f_0(x_0, x_1, x_2) + \cdots + \lambda_{r-1} f_{r-1}(x_0, x_1, x_2) = 0 \tag{3.7}$$

on a plane irreducible curve $F(x_0, x_1, x_2) = 0$.

We do not consider the fixed points of the system, i.e., the common roots of the basic polynomials $f_0, \ldots, f_{r-1}$, consequently all the divisors of the linear series change with $\lambda_0, \cdots, \lambda_{r-1}$.

Let $y_0, \ldots, y_{r-1}$ be homogeneous coordinates in the space P^{r-1} and let

$$\rho y_i = f_i(x_0, x_1, x_2). \tag{3.8}$$

When the point $x = (x_0, x_1, x_2)$ runs the plane curve F, the point $x = (y_0, \ldots, y_{r-1})$ cannot remain fixed since otherwise the functions f_i would differ by a constant factor and the linear series g_n^r has dimension 0, which case is not considered (we assume that the dimension of the linear system is exactly r, hence all the forms f_i are linearly independent). The point y runs a spatial curve C, consequently every point of the plane curve F corresponds to a unique point of the curve C whose coordinates are rational functions of the point x. Equalities (3.7) and (3.8) can be regarded as a definition of the spatial curve. A question arises when conversion is possible, i.e., when every point of C corresponds to a unique point of F.

Let x and x' be two points of the curve F converted to the same point y of the curve C.

Then

$$f_i(x) = \sigma f_i(x')$$

and every curve of system (3.7) passing through the point x also passes through the point x' related to x. The linear system possessing this property is called composite. Otherwise a system is simple. For a simple system the curves F and C are interrelated by a birational mapping, hence the coordinates of the point x also are rational functions of the coordinates of the point y.

Let a composite system enjoys the following property. Every divisor of the linear series which contains the point x also contains $\mu - 1$ other points changing with x. In this case the system is said to be order μ. For such a system every point of C corresponds to μ points of F, therefore, there is the algebraic interdependence $(1, \mu)$ between C and F.

An example of the simple system is the system of all curves of a given degree since for any two points x and x' one can always find a curve passing through x but not through x'.

An example of the composite system can be built up in the following way. Let the initial curve of degree n have a singular point P of multiplicity $n - 2$. Consider the system of curves of degree $(n - 3)$ such that P is a point of multiplicity $n - 3$. Every curve of the system breaks into $n - 3$ lines passing through P. Therefore, every curve of the system passing through the point Q_1 contains the entire line PQ_1. This is composite of order $\mu = 2$ (the canonical system of a hyperelliptic curve).

Consider the intersection of the spatial curve C with a hyperplane in P^{r-1}, given a simple system

$$\lambda_0 y_0 + \lambda_1 y_1 + \cdots + \lambda_{r-1} y_{r-1} = 0.$$

Clearly, the points of intersection correspond to a divisor of the linear series

$$\lambda_0 f_0(x_0, x_1, x_2) + \ldots + \lambda_{r-1} f_{r-1}(x_0, x_1, x_2) = 0$$

and conversely, every divisor of the series corresponds to a hyperplane. The degree of the divisor of intersection of the curve with a hyperplane is called the degree of the curve. Thus, for g_n^r the degree of the curve C is n.

If the series g_n^r is composite of order μ, then all n points of the divisor fall into groups containing μ points each. Every group converts to a single point of the curve C and the hyperplane intersects C at n/μ points. In this case the degree of the curve C is n/μ.

A point Q of a spatial curve of the degree s is said to have multiplicity m_Q, if the hyperplane passing through Q intersects the curve at $s - m_Q$ points different from Q. The genus of the curve C is defined as the genus of the plane curve which is in a birational correspondence with C.

We now consider the spatial curve C defined by the complete system $L(D)$. Such a curve is called normal. The dimension, degree, and genus of this curve are interrelated by the Riemann-Roch formula

$$r = s - g + 1 + i.$$

Here $s = \deg(D)$, $i = i(D)$. A curve is called special or not depending on whether or not $i \neq 0$.

Since $i(D) \leq g$, for any linear series the inequality

$$r \leq s + 1 \tag{3.9}$$

holds and the equality is attained only in the case of rational curve ($g = 0$).

Assume that the complete linear series g_n^r is composite of order μ. The projective model of this system is an algebraic curve of degree n/μ, therefore, by

(3.9) the inequality

$$r \leqslant (n / \mu) + 1$$

holds.

If the series is not special, then $r = n - g + 1$, therefore,

$$n \geqslant \mu(r - 1) = \mu(n - g).$$

The assumption $\mu \geqslant 2$ leads to the inequality

$$n \geqslant 2(n - g),$$

$$n \leqslant 2g.$$

Thus if $n \geqslant 2g + 1$, then the linear system is simple. The corresponding linear series has no fixed points, i..e.,

$$l(D - P) = l(D) - 1$$

for any point P (the corollary of the Riemann-Roch theorem).

Therefore, using the space $L(D)$ with deg $(D) \geqslant 2g + 1$ a curve can be birationally mapped onto the normal curve C lying in the space P^{r-1}, where

$$r = \deg(D) - g + 1.$$

The resulting curve C has no multiple points.

11. Differentials on a Curve

All the rational functions on a curve make up a field. Every field is an extension of a prime field (in the case of characteristic zero coinciding with the rational field Q and in the case of characteristic p coinciding with the residue field Z_p). The extension can be either algebraic or transcendental. The transcendence degree of a field extension K is the smallest value n such that K is algebraic over $k(x_1, ..., x_n)$ for some $x_1, ..., x_n \in K$. In this case K is said to be the field of algebraic functions of n variables over the field k.

If $n = 1$, then we deal with an algebraic curve: the functions on the curves make up a field of transcendence 1. Conversely, every such field can be put into correspondence with a curve whose functions make up the initial field. A point of the curve corresponds to the class of equivalent valuations of the field K.

The differentiation of the field K is the k-linear transformation

$$D: K \to K$$

such that

$$D(xy) = xD(y) + yD(x) \text{ for all } x,y \in K.$$

Let $F(X, Y)$ be the defining equation of a curve with the function field K. Since F is an irreducible polynomial, F is not divided by $F_Y = \partial F(X, Y)/\partial Y$ (we may assume that $F_Y \neq 0$). Obviously,

$$0 = d(F(x,y)) = F_X(x,y)dx + F_Y(x,y)dy,$$

hence

$$dy = udx,$$

$$u = -F_X(x,y)/F_Y(x,y).$$

Thus the differentials dy on the curve make up a one-dimensional vector space over the function field K.

For any polynomial $G \in k[X, Y]$ one can set

$$D(G) = G_X(x,y) - uG_Y(x,y).$$

This completes the definition of differentiation in the field of rational functions.

Let $Q_P(F)$ denote the set of functions on the curve F defined in P. The set $Q_P(F)$ is called the local ring at the point P. There exists an element t (the local parameter at the point P) such that any function of $Q_P(F)$ can be expressed as $f = ut^n$, where u is the unit of the ring (the invertible element). Any straight line $t = aX + bY + c$ passing P, which is not a tangent to F, can be chosen as t. The number n is independent of the choice of t and coincides with the multiplicity of a zero (pole) of the function f.

Every function in a neighbourhood of the point P can be expanded into the

Laurent series

$$\zeta_P(f) = \sum_{i=n}^{\infty} c_i t^i.$$

The set $\{\zeta_P(f)\}$ of such series over all the points of the curve is the adele of a function, or the distribution. The coefficient $\text{Res}_P(f)$ of t^{-1} is the residue of a function at P.

The differential form is the formal expression

$$\omega = u \, dv,$$

where u and v are arbitrary functions,

$$u = \sum a_\nu t^\nu, \quad v = \sum b_\nu t^\nu.$$

We find the formal derivative

$$\frac{dv}{dt} = \sum \nu b_\nu t^{\nu - 1}$$

and the product of series

$$u(t)\frac{dv(t)}{dt} = \zeta_P(t).$$

When P runs the points of the curve, we obtain the adele of the differential $\{\zeta_P(\omega)\}$. This adele defines the divisor of the differential

$$(\omega) = \sum \text{ord}_P[\zeta_P(\omega)]P$$

and the residue vector of the differential

$$(\text{Res}_{P_1}(\omega), \text{Res}_{P_2}(\omega), \cdots)$$

independent of the choice of local parameters.

For any differential we have

$$\deg((\omega)) = 2g - 2.$$

The differential which has no poles is called the differential of the first kind. Every such differential is identified with a ϕ-curve, so the divisor of the differential of the first kind coincides with the canonical divisor.

For any differential the sum of residues is 0:

$$\sum_P \text{Res}_P(\omega) = 0.$$

Let $\{\zeta_P(f)\}$ be the adele of the function and let $\{\eta_P(\omega)\}$ be the adele of the differential ω. There is the pairing (inner product) of f and ω

$$(f, \omega) = \sum \text{Res}(\zeta_P \eta_P).$$

Using (f, ω) one can establish the duality on curves. Denote by $\Omega(P)$ the space of differentials such that $(\omega) \geqslant D$. Then the specialization index $i(D) = \dim \Omega(W - D)$, where W is the canonical divisor. The Riemann-Roch theorem takes the form

$$l(D) = \deg(D) - g + 1 + l(W - D). \tag{3.10}$$

Example. Let F be the elliptic curve

$$x_0^3 + x_1^3 + x_2^3 = 0$$

over the field $F_4 = \{0, 1, \alpha, \beta\}$.

Here

$$\alpha^2 = \beta, \quad \beta^2 = \alpha, \quad \alpha + \beta = 1, \quad \alpha\beta = 1.$$

Consider the conic

$$\phi_0 = x_0 x_1 + x_1 x_2 + x_2 x_0$$

and denote by G the divisor of intersection of this conic with F. Obviously, $s = \deg(G) = 6$. By the Riemann-Roch theorem

$$l(G) = \deg(G) - g + 1 = 6.$$

We construct a basis of $L(G)$ using curves of degree 3 (in this case every curve is adjoint to F since F is nonsingular).

Choose a straight line L. The curve $\phi_0 L$ intersects F over the divisor of degree 3 in addition to the divisor G. All the curves of degree 3 passing through this remainder divisor make up the space $L(G)$. Obviously, curves ϕL, where ϕ

is a conic, enter this set. All such curves make up the space P^5 and since $l(G) = 6$, those are the only curves in the space $L(G)$. Thus the basis of $L(G)$ is made up of the conics

$$\{1, x_0^2 / \phi_0, x_1^2 / \phi_0, x_2^2 / \phi_0, x_0 x_1 / \phi_0, x_1 x_2 / \phi_0\}$$

The curve F contains 9 F_4-points

	P_1	P_2	P_3	P_4	P_5	P_6	P_7	P_8	P_9
x_0	1	0	α	1	0	β	0	α	1
x_1	0	1	0	1	α	0	β	1	α
x_2	1	1	1	0	1	1	1	0	0

The embedding of F in the space P^5 defined by the linear series $L(G)$ is of the form

	P_1	P_2	P_3	P_4	P_5	P_6	P_7	P_8	P_9
x_0^2	1	0	α	1	0	α	0	β	1
x_1^1	0	1	0	1	β	0	α	1	β
x_2^2	1	1	1	0	1	1	1	0	0
$x_0 x_1$	0	0	0	1	0	0	0	α	α
$x_1 x_2$	0	1	0	0	α	0	β	0	0
$x_0 x_2$	1	0	α	0	0	β	0	0	0

On dividing all coordinates by $\phi_0(P_j)$, we obtain the matrix in nonhomogeneous coordinates, which is reduced by linear transformations of rows to the canonical form

$$H = \left\{ \begin{matrix} 1 & & & & & \alpha & \alpha & 0 \\ & 1 & & & & \alpha & \beta & 1 \\ & & 1 & & & \beta & 0 & \alpha \\ & & & 1 & & 0 & 1 & 1 \\ & & & & 1 & \beta & \beta & 1 \\ & & & & & 1 & 1 & \alpha & \alpha \end{matrix} \right.$$

The dual matrix to H is

$$
\begin{bmatrix}
\alpha & \alpha & \beta & 0 & \beta & 1 & 1 & 0 & 0 \\
\alpha & \beta & 0 & 1 & \beta & \alpha & 0 & 1 & 0 \\
0 & 1 & \alpha & 1 & 1 & \alpha & 0 & 0 & 1
\end{bmatrix}
$$

The duality on curves show that every row of this matrix is the residue vector of the differential ω of the space $\Omega(G - D)$, where D is the divisor composed of all points of hte curve:

$$
D = P_1 + P_2 + \cdots + P_9
$$

hence

$$
(\mathrm{Res}_{P_1}(\omega), \mathrm{Res}_{P_2}(\omega), \ldots, \mathrm{Res}_{P_9}(\omega)) =
$$

$$
= (\alpha, \alpha, \beta, 0, \beta, 1, 1, 0, 0)
$$

for a differential $\omega \in \Omega(G - D)$. All these differentials make up a space of dimension 3 which results from all linear combinations of the rows.

12. Zeta Function of a Curve

In the element number theory, the Euler formula

$$
\prod_P \frac{1}{1 - 1/P} = \sum_{n=1}^{\infty} (1/n)
$$

is well known, where the product on the left is taken over all prime p.

The function

$$
\zeta(s) = \sum_{n=1}^{\infty} \frac{1}{n^s} = \prod_P \frac{1}{1 - \dfrac{1}{p^s}}
$$

of the complex variable $s > 1$ is called the zeta function. The Euler identity follows from the uniqueness of factorization into primes.

If we write an arbitrary effective divisor $D = \sum n_P P$ on a curve in the multiplicative form $D = \prod P^{n_P}$ and assume that P is 'a prime divisor', while D is an arbitrary 'integer divisor', then the unique factorization into primes is still valid. If the initial curve is defined over the field F_q and if the coordinates of the pint P lie in an extension $F_q m$ (but do not lie in any smaller one), then the point P is said to be of degree $m = \deg p$.

The number $q^{\deg P}$ is called the norm N_P of a prime divisor P. For an arbitrary effective divisor D, the norm is defined in such a way that the multiplicativity is preserved.

$$N_{D_1 D_2} = N_{D_1} N_{D_2}$$

Thus, if $D = \sum n_P P$, then

$$N_D = \prod q^{n_P \deg P}$$

We introduce the zeta function in the following manner

$$\zeta(s) = \prod_P \frac{1}{1 - \dfrac{1}{N_P^s}},$$

where the product is taken over all prime divisors P.

We change the variables as $u = q^{-s}$ and denote

$$Z(u) = \zeta(s),$$

and

$$\frac{d(\log Z(u))}{du} = \sum_{m=1}^{\infty} N_m u^m = \sum_P \sum_{\nu=1}^{\infty} \deg p u^{\nu \deg p}$$

hence

$$N_m = \sum_{\deg_p / m} \deg_p$$

Thus N_m is equal to the number of prime divisors of degree 1 in the extension $F_q m$, i.e., coincides with the number of $F_q m$-points of the curve.

On the other hand

$$\zeta(s) = \prod_P (1 + \frac{1}{N_P^s} + \frac{1}{N_P^{2s}} + \cdots) =$$

$$= \sum(1/N_{P_1}^{n_1 s} N_{P_2}^{n_2 s} \cdots N_{P_k}^{n_k s}) = \sum_D (1/N_D^s) = \sum_{n=0}^{\infty} I_n q^{-ns}$$

where the sum on the right is taken over all integer (effective) divisors, while I_n is equal to the number of integer divisors of degree n. Consequently, one can obtain exhaustive information on the number of the points of the curve by finding the number of integer divisors of different degrees.

For example, let $g = 0$ (a projective curve). In this case all the divisors of the same degree are linearly equivalent. The class of integer divisors of degree n is identified with the set of all polynomials of degree n and the number of such polynomials is

$$(q^{n-1} - 1)/(q - 1)$$

(two polynomials differing by a constant multiplier leads to the same divisor).

Since the norm of the divisor is q^n, we obtain the expression

$$\zeta(s) = \sum_{n=0}^{\infty} [(q^{n+1} - 1)/(q - 1)]q^{-ns} = 1/(1 - q^{-s})(1 - q^{1-s}).$$

Now let $g > 0$. We combine all the divisors involved in the summation into classes (relative to linear equivalence). Denote by h the number of divisor classes. All the divisors of the same class A are of the same degree which is called the class degree $\deg(A)$. We now separately consider the classes which have $\deg(A) \geqslant 2g - 1$ and the classes with $\deg(A) < 2g - 1$.

The Riemann-Roch theorem aims to determine the number of integer divisors lying in a class, hence in the first case we have

$$\dim(A) = n - g + 1$$

and the class contains

$$(q^{n-g-1} - 1)/(q - 1)$$

integer divisors.

For a fixed degree n, there exist exactly h classes. Therefore, the first sum is equal to

$$\zeta_1(s) = \sum_{n=2g-1}^{\infty} \frac{hq^{n-g-1} - 1}{q - 1} q^{-ns} =$$

$$= \frac{h}{q-1} \frac{q^{1-g+(2g-1)(1-s)}}{1-q^{(1-s)}} - \frac{h}{q-1} \sum_{n=2g-1}^{\infty} q^{-ns}$$

The second sum is

$$\varsigma_2(s) = \sum_{n=0}^{2g-2} \sum_{\deg A = n} \frac{q^{\dim(A)} - 1}{q-1} q^{-ns} =$$

$$= \frac{1}{q-1} \sum_{n=0}^{2g-2} \sum_{\deg (A)=n} q^{\dim(A)-ns} - \frac{h}{q-1} \sum_{n=0}^{2g-2} q^{-ns}.$$

On combining those two sums, we obtain

$$\varsigma(s) = \varsigma_1(s) + \varsigma_2(s) = \frac{1}{q-1} \sum_{n=0}^{2g-2} \sum_{\deg A = n} q^{\dim(A)-ns} +$$

$$+ \frac{h}{q-1} \Big(\frac{q^{1-g+(2g-1)(1-s)}}{q^{1-s}} + \frac{1-q^s}{1-q^s} \Big)$$

The latter expression implies that

$$\varsigma(s) = \frac{L(q^{-s})}{(1-q^{-s})(1-q^{1-s})},$$

(3.11)

where $L(u)$ is a polynomial with integer rational coefficients,

$$L(u) = 1 + (N_1 - q + 1)u + \cdots + q^g u^{2g}$$

It follows from the Riemann-Roch theorem that

$$q^{(g-1)}\varsigma(s) = q^{(g-1)(1-s)}\varsigma(1-s),$$

$$q^{gs} L(q^{-s}) = q^{g(1-s)} L(q^{s-1}).$$

On passing to the logarithmic derivative, we obtain

$$N_n = q^n + 1 - \sum_{\nu=1}^{2g} \omega_\nu^n$$

$$L(u) = \prod_{\nu=1}^{2g} (1 - \omega_\nu u)$$

$$\omega_{\nu-g} = q, \quad \nu = 1,\dots,g.$$

The values ω_ν possess the following property:

$$|\omega_\nu| = \sqrt{q}$$

which leads to the estimate of the number of F_q-points on the curve

$$|N - (q + 1)| \leqslant 2g\sqrt{q}$$

(the Hasse-Weil bound).

Comments

The Theory of linear series on algebraic curves is presented in the books of Severi [15], Semple and Roth [12], and Fulton [3]. The reader can get acquainted himself with the zeta function in the book of Eichler [1].

CHAPTER 4

Generalized Jacobian Codes

1. The $\mathfrak{M}$-equivalence on Curves. Definition of the Code.

Let X be an irreducible algebraic curve over the finite field F_q, $q = p^m$. Denote by S the finite set of the points of the curve X. If an integer $n_p > 0$ is specified for every $P \in S$, then the module with support S is said to be defined. Thus the module $\mathfrak{M}$ can be identified with the positive divisor $\sum n_P P$. Let g be a rational function on X. If

$$v_P(1 - g) \geqslant n_P, \quad P \in S, \tag{4.1}$$

then we use the notation

$$g \equiv 1 \mod \mathfrak{M}. \tag{4.2}$$

The set of all divisors on $X - S$ reduced modulo p can be identified with an affine space A^n over the field F_p, where $n = |X - S|$. The reduction modulo p means that in the divisor $\sum n_Q Q$ we replace n_Q by its remainder modulo p. In other terms this process can be described in the following way. Let U be the set of all divisors on $X - S$, i.e., those equal to zero outside this set.

Consider the endomorphism

$$\phi: U \to pU, \tag{4.3}$$

$$D \overset{\phi}{\to} pD.$$

The reduction of a divisor modulo p is equivalent to passing from U to U/pU so that $A^n \sim U/pU$.

We introduce the linear code C consisting of the vectors of the space A^n corresponding to the divisors (g) for all functions g, which satisfy conditions (4.2). Changing module $\mathfrak{M}$, we obtain a family of codes which are called the generalized Jacobian codes. This term is chosen because the error-checking matrix is connected to the generalized Jacobian of a curve. More precisely, the columns of this matrix correspond to the canonical embedding of a curve into its generalized Jacobian reduced modulo p, while the syndromes of the code coincide with the points of the variety.

Let G be the group of principle divisors, i.e., the divisors of functions g. Since $G \subseteq U$, it follows that the endomorphism ϕ induces the homomorphism

$$\psi: G \to U/pU \tag{4.4}$$

The image $\bar{G}$ under this homomorphism consists of principle divisors reduced modulo p.

According the well-known homomorphism theorem, this image is isomorphic to the groups

$$\bar{G} = G + pU/pU = G/(pU) \cap G = G/pG. \tag{4.5}$$

Here pG stands for the set of the principle divisors (g^p). Note that being reduced modulo p, the divisor of a function may no longer be principle.

We denote by B the group of principle divisors corresponding to the functions $g \equiv 1 \bmod \mathfrak{M}$. Obviously, $B \subseteq G \subseteq U$ and the reduction modulo p corresponds to the isomorphisms

$$\bar{B} = B/pB. \tag{4.6}$$

Here pB consists of the divisors of the functions g^p, where $g \equiv 1 \bmod \mathfrak{M}$. Note that the latter condition implies the fact that g is invertible at all points of the set S, i.e., has no zeros and poles in S. Obviously, the function g^p has the same property, hence pB do not intersect with $\mathfrak{M}$.

All the divisors of $\overline{B}$ make up in fact the linear code C. The quotient space $A^n / C = C^*$ corresponds to the dual code, i.e., the error-checking matrix. Clearly,

$$C^* = \overline{U} / \overline{B} \tag{4.7}$$

The divisors D and D' on $X - S$ are said to be $\mathfrak{M}$-equivalent, if

$$D - D' = (g) \in B.$$

This equivalence is denoted by

$$D \underset{\mathfrak{M}}{\sim} D'.$$

Also, D and D' are said to be $\mathfrak{M}^{(p)}$-equivalent, if

$$D - D' = (g) + pD'' \tag{4.8}$$

for a divisor D'',

$$g \equiv 1 \bmod \mathfrak{M}.$$

This equivalence is denoted by

$$D \underset{\mathfrak{M}^{(p)}}{\sim} D'.$$

Using these terms, one can calculate that the dual code C^* consists of the $\mathfrak{M}^{(p)}$-equivalence classes for the divisors on $X - S$.

2. Local Symbol. Error-Correcting Capacity of a Code

In Chapter 3, the differential was defined as a form expression $\omega = f du$ relating to two functions f and u. Moreover, the operation $\mathrm{Res}_t(\omega)$ was defined as a residue of the differential equal to the coefficient C_{-1} in the expansion of ω into the Laurent series at a point P with the local parameter t.

If t_1 and t_2 are two local parameters, then $\mathrm{Res}_{t_1}(\omega) = \mathrm{Res}_{t_2}(\omega)$. The operation $\mathrm{Res}_t(\omega)$ also possesses the following properties:

1) $\mathrm{Res}_t(\omega)$ is linear in ω;

2) $\mathrm{Res}_t(\omega) = 0$ for $v(\omega) \geqslant 0$, i.e., if P is not a pole of ω;

Of special importance are the 'logarithmic differentials' $\omega = dg\,/g$, defined for any nonzero function g. The following condition holds for these differentials: all zeros and poles of g are simple poles of ω, and the residues at those poles are equal to the multiplicity of a zero or a pole of the function (appropriately signed):

$$\mathrm{Res}_t(dg\,/g) = v(g).$$

Here $v(g)$ is an element of the prime field F_p, i.e., the multiplicities of zeros and poles of the function are reduced modulo p. Thus the code vector x can be defined as the vector of residues

$$x = (\mathrm{Res}_{P_1}(dg\,/g),...,\mathrm{Res}_{P_n}(dg\,/g))$$

at every point $P_i \in X - S$ of the function $g \equiv 1 \bmod \mathfrak{M}$. But for such functions we have

$$dg\,/g \equiv 0 \bmod \mathfrak{M} \tag{4.9}$$

if $\mathfrak{M} = p\mathfrak{M}'$, then all n_P in the modulo are divided by p. In what follows we shall assume that this condition holds true. Congruence (4.9) implies the estimate of the code separation

$$d \geqslant \deg(\mathfrak{M}) - (2g - 2). \tag{4.10}$$

Indeed, the Hamming weight of x coincides with the number of poles of the differential, and since $\deg((\omega)) = 2g - 2$, it follows that the number of zeros exceeds the number of poles by $2g - 2$.

We shall assume throughout that

$$\deg(\mathfrak{M}) \geqslant 2g - 2 \tag{4.11}$$

If two principle divisors D and D' are $\mathfrak{M}$-equivalent, then (4.8) implies that

$$D - D' = (gF^p),$$

where $g \equiv 1 \bmod \mathfrak{M}$ and F is invertible at the points of S. It follows that

$$D - D' = (g), \tag{4.12}$$

where $g \equiv F^p \bmod \mathfrak{M}$.

For this function we have

$$dg / g \equiv 0 \bmod \mathfrak{M}. \tag{4.13}$$

Conversely, congruence (13) implies (12).

3. The Structure of the Error-Checking matrix

The error-checking matrix consists of 2 blocks:

$$T = \begin{vmatrix} J^{(p)} \\ L \end{vmatrix}.$$

The first block 'cuts out' all divisors which are not principle, i.e., the zero space of the matrix $J^{(p)}$ consists of principle divisors. This block corresponds to the ordinary Jacobian of a curve and will be described in subsequent sections.

The second block singles out those divisors (g) among the principle divisors, which corersponds to the functions $g \equiv F^p \bmod \mathfrak{M}$. Consider the structure of matrix L.

If $\mathfrak{M} = \sum n_P P$, then for each $n_P = n$ the system of representatives in L is made up of the polynomials

$$f = a_0 + a_1 t + \ldots + a_{n-1} t^{n-1}$$

such that $a_0 \neq 0$, $a_i = 0$ for $i = p^j$, $j = 1, 2, \ldots$. In other words, we factorize by the set of polynomials F^p. It follows that

$$\dim L = ((p-1)/p) \deg(\mathfrak{M}). \tag{4.14}$$

This value is equal to the dimension of L over the field of definition F_q of the curve. If $q = p^m$, then over the prime field F_p, which is the field of definition of the code, this dimension satisfies the inequality

$$r_2 \leqslant m \frac{p-1}{p} \deg(\mathfrak{M}). \tag{4.15}$$

The total number of control symbols of the code is

$$r = r_1 + r_2,$$

where $r_1 = \dim J^{(p)}$.

In the specific case when the curve is rational ($g=0$), all the divisors are principle, therefore, $J^{(p)}$ reduces to a single row $(1,....,1)$ 'the even parity check'. Consequently, $r_1 = 1$ and the code is the rational code described in Chapter 1. When $p = 2$, this code has the parameters

$$d \geqslant 2t + 2, \quad n \leqslant 2^m + 1, \quad r \geqslant mt + 1.$$

The codes which are based on an elliptic curve ($g=1$), are described in the next section.

4. Elliptic Curves

Every curve of genus 1 can be represented in normal form

$$y^2 - y + \alpha xy = x^3, \quad \alpha^3 \neq 27. \tag{4.16}$$

Here α belongs to the field of definition of the curve. The modular invariant is defined as

$$j = -\frac{\alpha^3(\alpha^3 + 24)^3}{\alpha^3 + 27}. \tag{4.17}$$

Two elliptic curves are isomorphic if and only if their invariants j coincide.
In the case $p = 2$, (4.17) takes the form

$$j = \frac{\alpha^{12}}{1 + \alpha^3}.$$

Thus there exists a single class of isomorphic curves over the field F_2. The defining equation of such a curve is

$$y^2 - y = x^3, \tag{4.18}$$

which correspnds to the value

$$j = \alpha = 0.$$

We now study the structure of the divisor class group $Cl(X)$ on an elliptic curve.

All equivalent divisors on a smooth projective curve are of the same degree. Therefore, we can speak about the divisor class degree. The image of the homomorphism

$$\deg : Cl(X) \rightarrow Z,$$

is the whole group Z, while the kernel consists of zero degree classes $Cl^0(X)$. On a rational curve, all the divisors of the same degree are equivalent, hence $Cl^0(X) = 0$. The converse is also true: if $Cl^0(X) = 0$ on a curve, then this curve is rational. Indeed, in this case the divisor $P_1 - P_2$ is principle for any two points P_1 and P_2. Thus there exists a function f with a single pole P such that $l(P) \geqslant 2$. It follows from the Riemann-Roch theorem that $i(P) = g$. But if $g \neq 0$, the specialisation index of the divisor P is $g - 1$, so it is only possible that $g = 0$.

We choose an arbitrary point P_0 on an elliptic curve and get correspondence between every point $P \in X$ and the class C_P containing the divisor $P - P_0$. The mapping $P \rightarrow C_P$ defines a one-to-one correspondence between the points of the curve and classes $C \in Cl^0(X)$. In order to prove this fact, note that for two different points P_1 and P_2 the classes C_{P_1} and C_{P_2} are also different: if $C_{P_1} = C_{P_2}$, then $P_1 - P_0 \sim P_2 - P_0$, hence $P_1 \sim P_2$ and the curve X is rational.

We now show that in every class of zero degree there is a divisor of the form $P - P_0$. First, let D be any effective divisor. We prove that there exists a point $P \in X$ such that

$$D \sim P + lP_0. \tag{4.19}$$

If $\deg(D) = 1$, then (4.19) is true with $l = 0$. If $\deg(D) > 1$, then $D = D' + R$, $\deg(D') = \deg(D) - 1$, $D' > 0$. By induction, we may assume that (4.19) has been proved for D':

$$D' \sim Q + rP_0, \quad D \sim R + Q + rP_0.$$

If there is a point P such that

$$l(R + Q \sim P + P_0) \tag{4.20}$$

then (4.19) follows.

First let $R \neq Q$. We draw the line RQ through these points. This line intersects X at a third point S. Suppose that $S \neq P_0$ and draw the line SP_0. Then we have the third point of intersection P.

It is clear that

$$R + Q + S \sim S + P_0 + P$$

and (4.20) follows.

In the case when $R = Q$ or $S = P_0$ we draw a tangent to X instead of a chord.

Now let $\deg(D) = 0$. Then $D = D_1 - D_2$, $D_1 \geq 0$, $D_2 \geq 0$, $\deg(D_1) = \deg(D_2)$. On applying (4.19) to D_1 and D_2, we obtain $D_1 \sim R + lP_0$, $D_2 \sim P + lP_0$. By what we have just proved there exists a point Q satisfying (4.20), hence

$$D \sim R - P \sim P_0 - Q,$$

what we wished to prove.

Thus there is a one-to-one correspondence between the points of a smooth projective plane cubic curve and the elements of the group $Cl^0(X)$. Using this, one can extend the group operation of $Cl^0(X)$ to the set X itself. The resulting operation with the points of the curve is called addition. According to the definition, $P + Q = R$ whenever $C_P + C_Q = C_R$, i.e.,

$$P + Q \sim R + P_0.$$

Clearly, the point P_0 is a zero. The operation $+$ and the inversion of an element can be described in elementary geometrical terms. In order to find the sum $P + Q$, one needs to draw a line through P and Q, to find the point of intersection S with the curve, then to draw the line P_0S; the point R of intersection of this line with X is the desired sum $P + Q$.

The chords should be replaced by tangents whenever points coincide.

If a curve is defined over rationals Q, the resulting abelian group is finite-generated (the Mordell theorem), i.e., having a finite number of points of a curve, one can derive all the remaining points which are rational over Q by use of the above chord and tangent technique.

In the case of characteristic p, the divisor class group is finite abelian. In this group there may be no points of order p. In such a case the curve is called supersingular. For small values of $p = 2,3$, only those curves are supersingular whose modular invariant j takes on zero. For example, for $q = 2$ any curve is supersingular.

A code associated with a supersingular curve is called regular. Since the reduction of divisors modulo p entails the reduction of group $Cl^0(X)$ modulo p, it follows that for regular codes this group reduces to zero and the first block $J^{(p)}$ in the error-checking matrix reduces to a single row $(1,1,...,1)$ as in the case of a rational curve. Thus the elliptic regular code has the parameters

$$r \leqslant m\left(\frac{p-1}{p}\deg(\mathfrak{M})\right) + 1 \tag{4.21}$$

$$d \geqslant \deg(\mathfrak{M}).$$

If a curve is not supersingular, then the group of the points of order p in $Cl^0(X)$ is cyclic. In this case there is an additional generator in the block $J^{(p)}$. On eliminating $\deg(\mathfrak{M})$ in equations (4.21) and denoting $d = 2t + 2$, where t is the number of errors to be corrected, we obtain

$$r \leqslant m\frac{p-1}{p}(2t+2) + 1.$$

In the binary case we have

$$r \leqslant mt + m + 1.$$

Thus this estimate shows that the number of control symbols increases by m as compared with the rational code. Since the code length can be increased as compared with a rational curve, it follows that the elliptic codes prove to be more efficient than the rational ones in a certain range of transmission speed k/n.

5. Jacobian of a Curve

In the previous section it was shown that the structure of an abelian group iso-morphic to the divisor class group can be introduced on an elliptic curve. On an arbitrary curve of genus g, a similar group structure can be introduced on the symmetric product $X^{(g)}$ of g replicas of the curve. Every component of this pro-duct can be identified with an effective divisor of degree g.

Let D be a rational divisor on X, i.e., a divisor invariant under the Frobenius transformation $P \rightarrow \sigma P$ of points of the curve. This divisor contains the conjugate $n_p \sigma P$ along with every point $n_p P$. One can always choose a basis in the space $L(D)$ and $\Omega(D)$, which is rational over the field of definition of the curve. Furth-ermore, let P be an arbitrary point of a curve. If D is a special divisor, i.e., $i(D) > 0$, then

$$i(D + P) = i(D) - 1. \qquad (4.22)$$

Indeed, by the definition of the specialization index we have $i(D + P) \geqslant i(D) - 1$. Since $i(D) > 0$, it follows that there exists a nonzero differential ω of $\Omega(D)$ rational over k. Since P is an arbitrary point of the curve, we can assume that P is not a zero of ω. It follows that $\Omega(D + P) \neq \Omega(D)$, which proves (4.22).

Lemma. Let D be a rational divisor of degree 0 and let $M_1 ... M_g$ be arbitrary independent points. Then there exists a unique positive divisor Δ such that

$$\Delta \sim D + \sum_{i=1}^{g} M_i.$$

Proof. The specialization index satisfies the inequality $i(D) \leqslant g$. Writing the divisor $D + \sum_{i=1}^{g} M_i$ as

$$D + \sum_{i=1}^{g-1} M_i + M_g,$$

we obtain

$$i(D + \sum_{i=1}^{g} M_i) = i(D + \sum_{i=1}^{g-1} M_i) - 1,$$

whenever $i(D + \sum_{i=1}^{g-1} M_i) > 0$. Repeating this several times we come to the conclusion that $i(D + M) = 0$, where $M = \sum_{i=1}^{g} M_i$. It follows from the Riemann-Roch theorem that $l(D + M) = 1$, therefore there exists a unique positive divisor Δ satisfying the condition

$$\Delta \sim D + M, \quad \deg(\Delta) = g.$$

In the specific case when $D = 0$, we obtain $\Delta = M$.

Thus the linear equivalence divisor classes are in a one-to-one correspondence with the points of the symmetric product $X^{(g)}$. The group structure of the divisor classes is extended to X in the following way. Choose a point P_0 rational over k. If M and N are two arbitrary independent points of $X^{(g)}$, then there exists a unique positive R such that

$$R \sim M + N - gP_0.$$

The point $R \in X^{(g)}$ is viewed as the sum of points of M and N. One can readily see that this addition is commutative and associative. Obviously, the point gP_0 plays the role of zero. If $g = 1$, then $X^{(g)} = X$ and the above addition operation coincides with addition of points on a cubic curve.

The symmetric product $X^{(g)}$ with the described abelian group structure is called the Jacobian variety or Jacobian. The fact that the Jacobian depends on g parameters manifests itself in that the Jacobian variety has dimension g. The curve X can be embedded in the Jacobian in the following way

$$\theta: X \to J(X)$$

$$(x) = Cl(x - P_0).$$

Here $J(X)$ stands for the Jacobian and the canonical mapping θ sends every point x of the curve of the equivalence class containing the divisor $x - P_0$. Since $x - P_0$ is a divisor of degree 0, it follows from the lemma that there exists a positive divisor Δ of degree g such that $\Delta \sim x - P_0 + M$. This divisor belongs to $X^{(g)}$ and defines the point $\theta(x)$ of the Jacobian. Different points x correspond to different points $\theta(x)$. Indeed, if $x \neq y$ and $\theta(x) = \theta(y)$, then

$$\Delta \sim x - P_0 + M,$$

$$\Delta \sim y - P_0 + M.$$

It follows that

$$x \sim y,$$

which is possible only if $g = 0$.

The mapping θ can be extended to a homomorphism of the divisor group in J in the following manner: every divisor $D = \sum n_P P$ on X is put into correspondence with a point of the Jacobian

$$\theta(D) = \sum n_P \theta(P),$$

where on the right we make use of the addition of points of the Jacobian. In other words, we find a unique divisor $\Delta \in X^{(g)}$ such that

$$\Delta \sim D - (\deg(D))P_0 + M.$$

The mapping θ is a surjection, i.e., every points of the Jacobian corresponds to a divisor and the kernel of the mapping consists of principle divisors (the Abel-Jacobi theorem).

The reduction of divisors modulo p implies the passage from J to $J^{(p)} = J / pJ$. A curve is called supersingular if the Jacobian has no points of order p, where p is the characteristic of the ground constant field. For such a curve defined over a finite field, the order $|J|$ of the Jacobian is not divided by p. The supersingularity criterion is formulated in terms of the Hasse-Witt matrix of an algebraic curve, which is the modular invariant in the specific case of elliptic curves.

The code corresponding to a supersingular curve is said to be regular. For such a code $J^{(p)}$ reduces to a single row, hence we obtain the following estimates for the code parameters:

$$r \leqslant m \frac{p-1}{p} \deg(\mathfrak{M}) + 1,$$

$$d \geqslant \deg(\mathfrak{M}) - (2g - 2),$$

which is equivalent to the inequality

$$r \leqslant 2m \frac{p-1}{p}(t+g) + 1.$$

For $p = 2$ we obtain

$$r \leqslant m(t+g) + 1.$$

If a curve is defined over a prime field, then $m = 1$

$$r \leqslant 2 \frac{p-1}{p}(t+g) + 1.$$

For small values of n, the codes over a rational curve prove to be the best, while for large values of n, those are the best which have a large value of g since those curves can have far more points than the rational ones.

If the initial curve is not supersingular, then the elements of order p of the Jacobian make up a cyclic group, hence one more generator is added to the error-checking matrix.

6. The generalized Jacobian variety

As we showed in the previous section, the divisor class group of an arbitrary algebraic curve of genus g is identified with the points of the symmetric product $X^{(g)}$. Consider the $\mathfrak{M}$-equivalence on a curve and the corresponding $\mathfrak{M}$-equivalent divisor class group. The elements of this group can be identified with the points of the symmetric product $X^{(\pi)}$, where π is defined as

$$\pi = g \quad \text{for } \mathfrak{M} = 0,$$

$$\pi = g + \deg(\mathfrak{M}) - 1 \quad \text{for } \mathfrak{M} \neq 0.$$

The value π is interpreted as the arithmetic (virtual) genus of a curve $X_{\mathfrak{M}}$ with singularities, which is constructed from X and the module $\mathfrak{M}$. In this case the support S of divisor $\mathfrak{M}$ shrinks to a singular point Q which is assigned the multiplicity

$$\delta_Q = \deg(\mathfrak{M}) - 1, \quad \deg(\mathfrak{M}) \geqslant 2,$$

$$X_{\mathfrak{M}} = (X - S) \cup \{Q\}.$$

The functions and differentials on the curve $X_{\mathfrak{M}}$ are defined so that the $\mathfrak{M}$-equivalence of divisors on the curve X becomes the usual linear equivalence on the curve $X_{\mathfrak{M}}$.

Therefore, the generalized Jacobian $J_{\mathfrak{M}}$ (the $\mathfrak{M}$-equivalence class group) on the curve X corresponds to the ordinary Jacobian of the curve $X_{\mathfrak{M}}$ of genus π. Using this, one can extend the group structure to the symmetric product $X^{(\pi)}$. If D is a rational divisor of degree 0 and $M_1, ..., M_\pi$ are arbitrary independent points, then there exists a unique positive divisor such that

$$\Delta_{\mathfrak{M}} D + \sum_{i=1}^{\pi} M_i.$$

If M and N are two independent arbitrary points of $X^{(\pi)}$, then the sum is defined as

$$R_{\mathfrak{M}} M + N - \pi P_0,$$

where P is a point rational over k.

The canonical embedding of the curve X into $J_{\mathfrak{M}}$ is defined in a similar way:

$$\theta: X \to J_{\mathfrak{M}},$$

$$\theta(x) = Cl_{\mathfrak{M}}(x - P_0).$$

$J_{\mathfrak{M}}$ is a variety of dimension π such that

$$J_{\mathfrak{M}} / J \cong L.$$

Thus J_m is an extension of the Jacobian based on the linear group.

This fact shows that the error-checking matrix consists of the blocks $J^{(p)}$ and L. Consequently, every column of the error-checking matrix can be identified with a point of the variety $J_{\mathfrak{M}}^{(p)}$ corresponding to the canonical embedding

$$T = (\theta(P_1), ..., \theta(P_n)).$$

An arbitrary divisor $D = \sum n_p P$ on $X - S$ (a vector of the space A^n) corresponds to the syndrome

$$\tilde{S} = \sum n_p \theta(P).$$

The decoding problem takes the form: given a point $\tilde{S}$ of the variety $J_{\mathfrak{R}}^{(\mathfrak{P})}$, one has to find the divisor $D = \sum n_p P$ with minimum number of nonzero coefficients n_p (the leader of the class $\tilde{S}$) in the corresponding $\mathfrak{M}$-equivalence class $\theta^{-1}(\tilde{S})$. Since the divisor D is identified with the differential dg / g, it follows that the leader of the class $\tilde{S}$ can be identified with the differential of this form having the minimum number of poles. Since the code separation d satisfies the inequality $d \geqslant \deg(\mathfrak{M}) - (2g - 2)$, it follows that all differentials such that the number of poles satisfies the condition $t \leqslant \frac{1}{2}\deg(\mathfrak{M}) - g$ correspond to different points $\tilde{S}$ (syndromes).

7. The Fermat and Hermite Curves

The codes associated with curves prove to be the more efficient the more is their length, i.e., the more F_q-points are contained in the initial curve. In this connection we describe some curves with a large number of points in this and the next sections.

The Fermat curve is defined by the equation

$$X^{\mathfrak{M}} + Y^{\mathfrak{M}} = Z^{\mathfrak{M}}.$$

This curve is nonsingular, whenever $\mathfrak{M}$ is not divided by the characteristic of field $k: \mathfrak{M} \neq 0 \pmod{p}$. Therefore the genus of the curve is

$$g = (\mathfrak{M} - 1)(\mathfrak{M} - 2) / 2.$$

For small values of $\mathfrak{M}$ there are the following known values for the number of points of the Fermat curve over the field F_q, $q = p^h$ (the table is taken from the memoir of Segre [10]).

	p	h	n
3	-1(mod 3)	0(mod 2)	$q + 1 + 2\sqrt{q}$
	1(mod 8)	$\neq$ 0(mod 4)	
4	5(mod 8)	2(mod 4)	$q + 1 + 6\sqrt{q}$
	3(mod 4)		
5	-1(mod 20)	0(mod 2)	
	± 2(mod 5)	$\neq$0(mod 4)	$q + 1 + 12\sqrt{q}$
		0(mod 4)	
		$\neq$0(mod 8)	
7	3,5(mod 7)	0(mod 6)	$q + 1 + 30\sqrt{q}$
	13,17,19,23(mod 24)	$\neq$0(mod 12)	

The table presents the cases when the Hasse-Weil upper bound is attained:

$$n = q + 1 + 2g\sqrt{q}.$$

The analysis of the Jacobian of the Fermat curve is based on the prime expansion discovered by Faddeev in [2].

Let $q = p^{2h}$, $\mathfrak{M} = \sqrt{q} + 1$. In the field F_q there exists an involutory automorphism $\tau: x \to \bar{x} = x^{\sqrt{q}}$, which leaves the subfield $F_{\sqrt{q}}$ fixed. The correspoding equation of the Fermat curve can be written as the Hermit form

$$X\bar{X} + Y\bar{Y} + Z\bar{Z} = 0.$$

These curves are called Hermitian. The genus of the curve is $g = (q - \sqrt{q})/2$ and the number of $F_{\sqrt{q}}$-points attains the Hasse-Weil upper bound. In order to show this, note that the equation

$$X^{\sqrt{q}-1} + Y^{\sqrt{q}+1} + Z^{\sqrt{q}+1} = 0, \tag{4.21}$$

can be regarded as that of the line

$$y_0 + y_1 + y_2 = 0,$$

over the field $F_{\sqrt{q}}$, since $\alpha^{\sqrt{q}+1} \in F_{\sqrt{q}}$ for any $\alpha \in F_q$. The number of points of this curve with three nonzero coordinates is $n_3 = \sqrt{q} + 1 - n_2$, where n_2 is the number of points of the curve with two nonzero coordinates. Since $n_2 = 3$, it follows that $n_3 = \sqrt{q} + 1 - 3$. Every value of the coordinate y_0 corresponds to $\sqrt{q} + 1$ values of X in equation (4.21). This remains valid for the pairs (y_1, Y) and (y_2, Z). Therefore, the number of points on a Hermitian curve is

$$n = (\sqrt{q} - 1)\frac{n_3(\sqrt{q} + 1)^3 + n_2(\sqrt{q} + 1)}{q - 1} = q\sqrt{q} + 1.$$

The Hermitian curves are invariant with respect to the projective unitary group $PGU(3, q)$ and every point has a tangent of multiplicity $\sqrt{q} + 1$.

8. Modular Curves

On a plane nonsingular curve, the number of F_q-points cannot exceed $q^2 + q + 1$. However, there exists nonsingular curves in the space P^N which have substantially more points. On projecting those curves on a plane, one obtains a curve with multiple F_q-points. For any preset N, there exist curves of genus g, which cannot be embedded in P^N in such a way that all F_q-points are nonsingular. Such curves were first constructed by Ihara [6] in connection with the zeta-function for modular curves.

A modular curve is defined by a modular group, i.e., the group $G = SL_2(Z)$ of integer second-order matrices with determinant 1. This group acts on the upper complex half-plane H through the mapping

$$\alpha(z) = (az + b) / (cz + d).$$

Let (ω_1, ω_2) be a pair of complex numbers such that $Im(\omega_1 / \omega_2) > 0$. Every such a pair correspnds to a lattice L:

$$L(\omega_1, \omega_2) = Z\omega_1 \oplus Z\omega_2,$$

i.e., the set of complex numbers of the form $m\omega_1 + n\omega_2$.

Let (ω_1', ω_2') be another pair generating the same lattice. Then

$$\omega_1' = a\omega_1 + b\omega_2$$

$$\omega_2' = c\omega_1 + d\omega_2,$$

with integer a, b, c, and d.

Denote $z = \omega_1 / \omega_2$, $z' = \omega_1' / \omega_2'$. We have

$$z' = (az + b)/(cz + d) = gz$$

$$g = \begin{bmatrix} a & b \\ c & d \end{bmatrix}.$$

Since (ω_1, ω_2) can also be represented as a linear integer combination of ω_1' and ω_2', it follows that $\det(g) = \pm 1$. The fact that $Im(\omega_1 / \omega_2)$ and $Im(\omega_1' / \omega_2')$ are of the same sign, implies

$$\det(g) = 1, \quad \text{so that } g \in G.$$

Thus two pairs (ω_1, ω_2) and (ω_1', ω_2') generate the same lattice if and only if they are congruent modulo $SL_2(Z)$. Therefore, the set of lattices in C can be identified with the set of orbits H / G under the action of G.

On the other hand, every lattice L is put into correspondence with the field of elliptic functions, i.e., the functions meromorphic on C which are L-periodic:

$$f(z + \omega) = f(z), \quad z \in C, \quad \omega \in L.$$

This field is generated by the Weierstrass functions $\rho(z)$ and $\rho'(z)$:

$$\rho(z) = \frac{1}{z^2} + \sum_{\omega \in L'} \left[\frac{1}{(z - \omega)^2} - \frac{1}{\omega^2} \right]$$

$$\rho'(z) = \frac{d\rho(z)}{dz} = -2 \sum_{\omega \in L'} \frac{1}{(z - \omega)^3}.$$

Here L' stands for the set of all nonzero periods of the lattice L.

The points $(\rho(z), \rho'(z))$ lie on the curve described by the equation

$$y^2 = 4x^3 - g_2 x - g_3. \tag{4.22}$$

The cubic polynomial on the right has the discriminant

$$\Delta = g_2^3 - 27 g_3^2 \neq 0.$$

Equation (4.22) defines an elliptic curve in the Weierstrass normal form. Any elliptic curve can also be represented in this form provided char $\neq$ 2, 3. In this case the coefficients g_2 and g_3 belong to the field of definition of a curve.

Thus every lattice L (hence the complex torus C / L) corresponds to an elliptic curve. The converse is also true, every elliptic curve is isomorphic to a torus C / L. Two tori C / L and C / M are isomorphic if and only if there exists a complex number such that $\alpha L = M$. In this case the lattices L and M are called equivalent. The equivalent lattices lead to isomorphic elliptic curves. Thus we obtain the following interpretation of the set of orbits H / G: this is the set of cosets of isomorphic elliptic curves. Every such a curve is defined uniquely up to an isomorphism by its modular invariant

$$j = 1728 g_2^3 / \Delta,$$

(in Section 4 we considered another invariant corresponding to the Deuring normal form which is suited for any characteristic including $p = 2$ and 3).

The function j defines a bijection (a one-to-one correspondence) between H / G and C. In order to stress the dependence of j upon the lattice, we write

$$j(z) = 1728 g_2^3(\omega_1, \omega_2) / \Delta(\omega_1, \omega_2),$$

$$z = \omega_1 / \omega_2.$$

We now consider the subgroup of $SL_2(Z)$ resulting from the reduction of integers modulo N. The natural homomorphism $Z \to Z_N = Z / NZ$ induces a homomorphism of matrices $SL_2(Z) \to SL_2(Z_N)$ such that the matrix

$$g = \begin{bmatrix} a & b \\ c & d \end{bmatrix},$$

corresponds to the matrix

$$\tilde{g} = \begin{bmatrix} \tilde{a} & \tilde{b} \\ \tilde{c} & \tilde{d} \end{bmatrix},$$

where $\tilde{a}$, $\tilde{b}$, $\tilde{c}$, and $\tilde{d}$ are residues modulo N. The kernel of this mapping is called the principle congruence subgroup of level N and consists of the matrices satisfying the condition

$$G_N = \{g \in SL_2(Z) | g \equiv 1 (\text{mod } N)\} = \left\{ \begin{bmatrix} a & b \\ c & d \end{bmatrix} \in SL_2(Z) | \ a \equiv \right.$$

$$d \equiv 1, \quad b \equiv c \equiv 0 \text{ mod } N\}.$$

G_N is the normal divisor of the group G, hence G / G_N is isomorphic to the group $SL_2(Z_\cdot)$. It follows that the index $\nu_N = 6$ when $N = 2$ and

$$\nu_N = [G : G_N] = \frac{N^3}{2} \prod_{p | N} (1 - \frac{1}{p^2}) \quad \text{if } N > 2.$$

Under the action of G, every orbit corresponds to ν_N orbits under G_N, this correspondence being algebraic, i.e., H / G_N is an algebraic curve that is a covering of degree ν_N of the projective curve $H / G = P^1$ over the field C. This curve is denoted by $X(N)$ and called modular. This curve has genus

$$g = \begin{cases} 0 & \text{if } N = 2, \\ 1 + \dfrac{N-6}{12N} \nu_N & \text{if } N > 2. \end{cases} \qquad (4.23)$$

The field of functions on the curve $X(N)$ is generated by two elements $(j(z), f_a(z))$, where f_a is represented through the Weierstrass function in the following manner

$$f_a(z) = \frac{g_2 g_3}{\Delta} \rho(a_1 z + a_2), \quad \omega_1 = z, \quad \omega_2 = 1,$$

where a_1 and a_2 are rational numbers of the form $\mathfrak{M} / N$, $\mathfrak{M} \neq 0 (\text{mod } N)$. The functions $j(z)$ and $f_a(z)$ admits the expansion in the Fourier series in powers of $q = e^{2\pi i z}$. On eliminating q from those expansion, we obtain an equation of degree ν_N of the form $\Phi(x, y) = 0$, where $x = j(z)$, $y = f_a(z)$. This equation has integer rational coefficients and thus can be reduced modulo a prime p. Therefore, we obtain a modular curve over the finite field F_p. Every point of this curve has the form $(j \bmod p, f_a \bmod p)$. Of particular interest are those values of $j \bmod p$, which correspond to the supersingular elliptic curves. Every such a point is called supersingular. This means that the x-coordinate of this point defines an elliptic curve with invariant j, which has no point of order p. If p is fixed and j is variable, then all the supersingular values of j make up a finite set

S lying in F_{p^2}. The cardinality of this set is

$$|S| = \begin{cases} 1 & p = 2, 3 \\ \dfrac{p-1}{12}, \dfrac{p+7}{12}, \dfrac{p+5}{12}, \dfrac{p+13}{12} & p = 1, 5, -5, -1, \bmod 12 \end{cases}$$

If $N \not\equiv 0 \pmod{p}$, then we have the following estimate for the number n of different supersingular points of the curve $X(N)$ with coordinates in the field F_{p^2}:

$$n \geq (g-1)(p-1).$$

Ihara [6] singled out a broad class of curves for which there is a similar estimate for the number of points.

Specific cases of the Ihara curves are the classical modular curves $X(N)$, the Shimura curves, and other curves related to the modular group.

9. Codes Based on Rational Mappings

Let H be the error-checking matrix of the Jacobian code. The columns of this matrix are numbered by the points of a curve, and since we consider curves over a finite field, it follows that every column can be represented in the form

$$\begin{pmatrix} f_1(P_i) \\ f_2(P_i) \\ \cdots \\ f_r(P_i \end{pmatrix},$$

where $f_j(P_i)$ is the value of a rational function f_j at the point $P_i \notin S$. Every such a function can be viewed as a mapping from the complement of S into the commutative group of the field F_q:

$$f: X - S \to F_q.$$

The mapping f can be linearly extended to a homomorphism of the group of divisors, which are zero on S, in F_q. In particular, if $g \equiv 1 \pmod{\mathfrak{M}}$, then the element $f((g))$ is defined as

$$f((g)) = \sum_{P \in X - S} v_P(g) f(P).$$

Since $(v_{P_1}(g), ..., v_{P_n}(g))$ is the code vector, it follows that $f((g))$ results from the multiplication of this vector by a row of the error-checking matrix, hence

$$f_i((g)) = 0, \quad i = 1, ..., r.$$

In this case $\mathfrak{M}$ is said to be the module of f_i (or f_i is associated with $\mathfrak{M}$).

The local symbol $(f, g)_P$ defined as

$$(f, g)_P = \operatorname{Res}_P(f \frac{dg}{g})$$

possesses the following properties

$$1) \; (f, gg')_P = (f, g)_P + (f, g')_P,$$

since

$$\frac{d(gg')}{gg'} = \frac{dg}{g} + \frac{dg'}{g'};$$

$$2) \; (f, g)_P = v_P(g) f(P) \quad \text{for} \quad P \in X - S.$$

Indeed, dg / g has a simple pole at P, which is also true for $f \frac{dg}{g}$ (since $P \notin S$) and, therefore, we can write

$$\operatorname{Res}_P(f \frac{dg}{g}) = f(P) \operatorname{Res}_P(\frac{dg}{g}) = f(P) v_P(g).$$

$$3) \; \sum_{P \in X} (f, g)_P = 0,$$

(this is the residue formula $\sum_{P \in \lambda} \operatorname{Res}_P \omega = 0$ applied to the differential

$$\omega = f \frac{dg}{g}).$$

We now assume that at every point $P \in S$ the function f has a pole of order no grater than n_P with $\mathfrak{M} = \sum n_P P$, $n_P \equiv 0 \pmod{p}$. In other words, we assume that $f \in L(\mathfrak{M})$. Then for $g \equiv 1 \pmod{\mathfrak{M}}$ we have

$$v_P(1-g) \geqslant n_P,$$

$$v_P(dg) \geqslant n_P \geqslant -v_P(f).$$

Since $v_P(g) = 0$, it follows that

$$v_P(f\frac{dg}{g}) \geqslant 0,$$

hence $\text{Res}_P(f\frac{dg}{g}) = 0$, so

4) $(f,g)_P = 0$ for $P \in S$ and for all functions $f \in L(\mathfrak{M})$.

in other words, every function of $L(\mathfrak{M})$ is associated with the modulo $\mathfrak{M}$, hence

$$f((g)) = 0 \text{ for } g \equiv 1 \pmod{\mathfrak{M}}, \ f \in L(\mathfrak{M}).$$

Let $\{f_1, f_2, ..., f_r\}$ be a basis of the space $L(\mathfrak{M})$. According to the Riemann-Roch theorem

$$r = \deg \mathfrak{M} - g + 1,$$

under the assumption that $\deg \mathfrak{M} \geqslant 2g - 1$. Moreover, suppose that $\deg \mathfrak{M} \geqslant 2g + 1$. Then the projective model of the linear series $L(\mathfrak{M})$ is defined by a normal nonsingular curve in the space P^{r-1}, every point of which corresponds to exactly one point of the initial plane curve

$$H = \begin{bmatrix} f_1(P_1) & f_1(P_2) & \cdots & f_1(P_n) \\ f_2(P_1) & f_2(P_2) & \cdots & f_2(P_n) \\ \cdots & \cdots & \cdots & \cdots \\ f_r(P_1) & f_r(P_2) & \cdots & f_r(P_n) \end{bmatrix}.$$

The degree of the spatial curve H is $\deg \mathfrak{M}$. If this curve is projected from a point $Q \in P^{r-1}$ on a hyperplane, then one obtains a curve of the degree

$$S = \begin{cases} \deg \mathfrak{M} & \text{if } Q \text{ lies outside the curve;} \\ \deg \mathfrak{M} - 1 & \text{if } Q \text{ lies on the curve.} \end{cases}$$

In the former case the curve resulting from projection is not normal in P^{r-2}, in the latter case the curve is normal in P^{r-2}.

Let $N \leqslant \deg \mathfrak{M} - 2g + 1$.

On projecting the curve consecutively from N points lying on the curve, we obtain a curve degree

$$s' = \deg \mathfrak{M} - N \geqslant 2g - 1,$$

which is normal in a space P^l. This curve is not special, therefore,

$$l = s' - g + 1 = \deg \mathfrak{M} - N - g + 1 = r - N.$$

This means that any N points of the curve are independent, i.e., they impose independent conditions on the hyperplanes in the space P^{r-1}. In other words, there exists a hyperplane passing through $N - 1$ points and not passing through the remaining point. Therefore, under projection from those points, the dimension reduces every time exactly by 1. Consequently, when $N \leqslant \deg \mathfrak{M} - 2g + 1$, every N points of the curve are independent and the code separation is equal at least to $N + 1$, hence

$$d \geqslant N + 1 \geqslant \deg \mathfrak{M} - 2g + 2.$$

Thus the codes defined by the projective model have the parameters

$$r = \deg \mathfrak{M} - g + 1,$$

$$d \geqslant \deg \mathfrak{M} - (2g - 2),$$

and

$$r \leqslant d + g - 1.$$

In Section 11 of Chapter 3 we considered an example of the Hermittian curve $X^3 + Y^3 + Z^3 = 0$ over the field F_4. The divisor of intersection of the curve with the conic

$$XY + YZ + ZX = 0,$$

was taken as the modulo $\mathfrak{M}$, hence $\deg \mathfrak{M} = 6$, $g = 1$, $n = 9$. The corresponding rational mapping of $L(\mathfrak{M})$ leads to the curve of degree 6 in P^5.

The code has the parameters $r = 6$, $d = 6$.

We now consider another example. The curve

$$F = Y^4 Z \quad YZ^4 + X^5 - X^2 Z^3,$$

over the field F_4 is a nonsingular quintic passing through the entire affine plane A^2 and a point at infinity, hence $q = 4$, $g = 6$, $n = 17$.

We choose the point $P = (0,1,0)$ on the curve and the module $\mathfrak{M} = 16P$. In order to find a basis of the space $L(\mathfrak{M})$, we make use of the Brill-Noether method. In this case the adjoint curves are all of the same degree. The point P has the tangent $Z = 0$ since at this point

$$F_X = X^4 |_P = 0, \quad F_Y = Z^4 |_P = 0,$$

$$F_Z = Y^4 - X^2 Z^2 |_P = 1,$$

and the equation of the tangent at the point (a,b,c) is

$$F_X(X - a) + F_Y(Y - b) + F_Z(Z - c) = 0.$$

This tangent intersects with F over the divisor $ZF = 5P$, i.e., has no points of intersection different from P. The line $X = 0$ intersects with F over the divisor

$$XF = P + \begin{pmatrix} 0 \\ 0 \\ 1 \end{pmatrix} + \begin{pmatrix} 0 \\ 1 \\ 1 \end{pmatrix} + \begin{pmatrix} 0 \\ \alpha \\ 1 \end{pmatrix} + \begin{pmatrix} 0 \\ \beta \\ 1 \end{pmatrix}.$$

Consequently, the curve $Z^3 X$ intersects with F over the remainder divisor consisting of 4 points in addition to the divisor $\mathfrak{M} = 16P$.

All the curves of degree 4 make up the space P^{14}:

$$a_1 X^4 + a_2 Y^4 + a_3 Z^4 + a_4 X^3 Y + a_5 Y^3 X +$$

$$a_6 X^3 Z + a_7 Z^3 X + a_8 X^2 Y^2 + a_9 X^2 Z^2 + a_{10} Y^2 Z^2 +$$

$$a_{11} Y^3 Z + a_{12} Z^3 Y + a_{13} X^2 YZ + a_{14} Y^2 XZ + a_{15} Z^2 XY.$$

The curves passing through $(0,0,1)$ are specified by the condition $a_3 = 0$.

The curves passing through the point $(0,1,1)$ are specified by the condition

$$a_2 + a_3 + a_{10} + a_{11} + a_{12} = 0.$$

Similarly, we have

$$a_2\alpha + a_3 + a_{10}\beta + a_{11} + a_{12}\alpha = 0,$$

$$a_2\beta + a_3 + a_{10}\beta + a_{11} + a_{12}\beta = 0.$$

These conditions lead to the equations

$$a_3 = a_{11} = a_{10} = 0; \quad a_2 = a_{12},$$

hence the basis of $L(\mathfrak{M})$ is made up of the curves

$$\{X^4, Y^4 + YZ^3, X^3Y, Y^3X, X^3Z, Z^3X, X^2Y^2,$$

$$X^2Z^2, X^2YZ, Y^2XZ, Z^2XY\},$$

$$r = 16 - 6 + 1 = 11.$$

On dividing all the curves by Z^3X, we obtain the following in nonhomogeneous coordinates:

$$\{X^3/Z^2, (Y^4 + YZ^3)/Z^3X, X^2Y/Z^3, Y^3/Z^3,$$

$$X^2/Z^2, 1, XY^2/Z^3, X/Z, XY/Z^2, Y^2/Z^2, Y/Z\}.$$

We anticipate to get a code with parameters

$$r = 11, d = \deg \mathfrak{M} - (2g - 2) = 16 - 10 = 6,$$

but encounter certain problems.

First of all, in order to evaluate the function $(Y^4 + YZ^3)/Z^3X$, note that

$$\frac{Y^4 + YZ^3}{Z^3X} \equiv \frac{XZ^3 + X^4}{Z^4} \pmod{F},$$

since

$$(Y^4 + YZ^3)Z - (XZ^3 + X^4)X = F.$$

But the function $(XZ^3 + X^4)/Z^4$ is zero at all the points of the curve which are rational over F_4, hence in this field $X^4 = X$. Consequently, we can omit this function and obtain the estimate $r \leqslant 10$ for the number of control symbols rather than the anticipated estimate $r \leqslant 11$.

Thus instead of $L(16P)$ we can consider $L(15P)$. Furthermore, on constructing the error-checking matrix and reducing it to the canonical form, we obtain the following code

$$
\begin{array}{ccccccccccc}
1 & 1 & 0 & 1 & 1 & 0 & 0 & 1 & 1 & 1 & 1 \\
1 & 1 & \beta & \beta & \beta & 0 & \alpha & \alpha & 0 & \alpha & 0 & 1 \\
1 & 0 & \alpha & \beta & \alpha & 0 & \beta & \alpha & 1 & \beta & 0 & 1 \\
1 & 1 & 1 & 1 & 0 & 1 & 1 & 0 & 0 & 1 & 0 & 1 \\
0 & 1 & \alpha & \beta & \alpha & 1 & \alpha & \beta & 0 & \beta & 0 & 1 \\
1 & 1 & \beta & \alpha & \beta & 1 & \beta & \beta & 1 & \beta & 0 & 1
\end{array}
$$

It is easily seen that the code separation is $d = 8$ instead of anticipated $d = 6$.

In order to understand the reason for this phenomenon, we find the number of poles of the basic functions of $L(\mathfrak{M})$.

The function Y/Z has 5 poles, since $Y = 0$ does pass through P. The function X/Z has 4 poles since X does not pass through P with mulitplicity 1. Therefore, X^2/Z^2 has 8 poles and Y^2/Z^2 has 10 poles. Similarly, we find the number of poles for other functions:

$$
\begin{array}{cccccccccccccccccc}
0 & 1 & 2 & 3 & 4 & 5 & 6 & 7 & 8 & 9 & 10 & 11 & 12 & 13 & 14 & 15 & 16 \\
+ & - & - & - & + & + & - & - & + & + & + & - & + & + & + & + & +
\end{array}
$$

No.	Number of poles		1	2	3	4	5	6	7	8	9	10	11	12	13	14	15	16
1	12	X^3/Z^3	0	0	0	0	1	1	1	1	1	1	1	1	1	1	1	1
2	13	X^2Y/Z^3	0	0	0	0	0	1	α	β	0	β	1	α	0	α	β	1
3	15	Y^3/Z^3	0	1	1	1	0	1	1	1	0	1	1	1	0	1	1	1
4	8	X^2/Z^2	0	0	0	0	1	1	1	1	β	β	β	β	α	α	α	α
5	0	1	1	1	1	1	1	1	1	1	1	1	1	1	1	1	1	1
6	14	XY^2/Z^3	0	0	0	0	0	1	β	α	0	α	1	β	0	β	α	1
7	4	X/Z	0	0	0	0	1	1	1	1	α	α	α	α	β	β	β	β
8	9	XY/Z^2	0	0	0	0	0	1	α	β	0	α	β	1	0	β	1	α
9	10	Y^2/Z^2	0	1	β	α	0	1	β	α	0	1	β	α	0	1	β	α
10	5	Y/Z	0	1	α	β	0	1	α	β	0	1	α	β	0	1	α	β
		X	0	0	0	0	1	1	1	1	α	α	α	α	β	β	β	β
		Y	0	1	α	β	0	1	α	β	0	1	α	β	0	1	α	β
		Z	1	1	1	1	1	1	1	1	1	1	1	1	1	1	1	1

Figure 5.

If P is an arbitrary point on a nonsingular curve F of genus g, then we denote

$$N_M = N_m(P) = l(mP).$$

It is clear that

$$1 = N_0 \leqslant N_1 \leqslant \dots \leqslant N_{2g-1} = g.$$

Thus there exists exactly g numbers

$$0 < n_1 < n_2 \dots < n_g < 2g,$$

such that there is no function with pole at the point P whose order is exactly n_i. These values n_i are called the Weierstrass gaps and $(n_1,\dots,n_g)$ is the sequence of gaps at P.

A typical sequence is $(1,2,\dots,g)$ and if the sequence of gaps has a different form, then P is called the Weierstrass point.

Every curve has only a finite number of the Weierstrass points and for $g > 1$ there exists at least one such a point.

The following conditions are equivalent:

a) P is a Weierstrass point;

b) $l(gP) > 1$;

c) $i(gP) > 0$. Indeed,

$$l(gP) = g - g + 1 + i(gP).$$

If P is not a Weierstrass point, then $l(gP) = 1$ by definition, hence $i(gP) = 0$ and there is no canonical divisor passing through gP. But if P is a Weierstrass point, then $l(gP) > 1$ and $i(gP) > 0$. In this case the divisor gP is special.

If r and s are not gaps, the $r + s$ also is not a gap. It suffices to construct the function $f_{r-s} = f_r f_s$ in order to show this.

The number n is a gap at P if and only if there exists a differential of the first kind ω such that $\mathrm{ord}_P(\omega) = n - 1$.

Returning to our example, we see that the point $P = (0,1,0)$ is Weierstrass with the sequence of gaps

$$(1,2,3,6,7,11).$$

Note that $11 = 2g - 1$ is a gap since there exists the canonical divisor $w = 10P = (2g - 2)P$ which is the divisor of intersection of the conic Z^2 with a curve, in this case all effective canonical divisors are cut out by curves of second order.

By virtue of duality on curves, every code vector is the vector of residues of a differential

$$\mathrm{Res}_{P_1}(\omega), \ldots, \mathrm{Res}_{P_n}(\omega)),$$

where ω can have poles only at points $P_1, \ldots, P_n$, every pole is simple, and ω passes through $\mathfrak{M}$. Consequently, $\mathrm{ord}_P(\omega) \geqslant 16$.

If the estimate of the code separation were tight, i.e., $d = 6$ then the divisor of zeros of the differential ω would be equal to $16P$. All the canonical divisors make up an equivalence class, therefore, on subtracting the canonical divisor $10P$ from $16P$, we would obtain a divisor of zeros of a function. However, $6P$ cannot be a divisor of zeros of a function since it is not a divisor of poles (the number 6 is a gap at P). Consequeltly, the code separation is greater than 6. We now clear up whether this number can be equal to 7.

Since $l(6P) = 3$, it follows that $i(6P) = 2$, i.e., there exist two independent conics passing through $6P$. The first conic is Z^2 and the second one is ZX. Suppose that for a Q the divisor $6P + Q$ is a divisor of poles of a function. Then $l(6P+Q) = 4$ and $i(6P+Q) = 2$. This means that the two conics Z^2 and ZX pass through the point Q in addition to $6P$. So we have come to a contradiction since Z^2 and ZX intersect one another over the divisor $6P$. Thus the divisor $6P + Q$ cannot be a divisor of zeros (poles) of a function and it follows that the code separation is at least 8.

In order to obtain the binary code from the constructed 4-adic one, one should pass to the subcode over a subfield. The error-checking matrix of such a code is of the form

$$H' = \begin{bmatrix} H \\ H^\sigma \end{bmatrix},$$

where H^σ results from H after squaring each of the rows. If some rows (functions) in the initial matrix H are conjugate, then they lead to the same rows of the matrix H'. For example, it suffices to take only one function Y/Z or Y^2/Z^2, say Y/Z, while the other one is superfluous. Similarly, X/Z and X^2/Z^2 are conugate. The functions X^3/Z^3 and Y^3/Z^3 take their values in the field F_4, therefore, their images in H^σ are redundant. Finally, the functions $(X^2Y/Z^3)^2$ and XY^2/Z^3 take on equal values at all points of the curve, therefore, one of these functions is superfluous. Denoting

$$0 = \binom{0}{0}, \quad 1 = \binom{0}{1}, \quad \alpha = \binom{1}{0}, \quad \beta = \binom{1}{1},$$

we obtain the matrix of the binary code (fig. 6).

After reducing this to the canonical form, we find the generating matrix

$$
\begin{array}{ccccccccccccc}
1 & 0 & 1 & 1 & 1 & 1 & 1 & 0 & 0 & 0 & 1 & 1 \\
1 & 1 & 0 & 1 & 1 & 0 & 0 & 0 & 1 & 1 & 1 & 1 \\
1 & 1 & 1 & 0 & 1 & 0 & 1 & 1 & 0 & 1 & 0 & 1 \\
1 & 1 & 1 & 1 & 0 & 1 & 0 & 1 & 1 & 0 & 0 & 1 \\
 & 1 & 0 & 0 & 0 & 0 & 1 & 1 & 1 & 1 & 1 & 1 & 1 \\
\end{array}
$$

The code has the parameters

$$n = 16, \quad r = 11, \quad d = 8.$$

This code coincides with the normal rational code over F_{16} with parametrization $\{1, t, t^2, ..., t^7\}$.

Moreover, the spatial curve over F_4 underlying the code coincides with the Del Pezzo affine surface of degree 9 over F_4 which is normal in P^9, since the curves of the linear series $L(15P)$ make up the Veronese variety of all curves of degree 3.

This thoroughly studied example shows that the estimates of the code parameters, which are valid for all curves and the corresponding codes, can be improved in some specific cases.

On the Fermat curve

$$X^l + Y^l = Z^l,$$

the point $P = (0,1,1)$ is the Weierstrass point for which only values $(0, l-1, l, 2l-2, 2l-1, 2l, 3l-3, ...)$ are not gaps with the divisor $W = (l-3)lP$ is canonical.

	1	2	3	4	5	6	7	8	9	10	11	12	13	14	15	16
X/Z	0	0	0	0	0	0	0	0	1	1	1	1	1	1	1	1
X/Z	0	0	0	0	1	1	1	1	0	0	0	0	1	1	1	1
Y/Z	0	0	1	1	0	0	1	1	0	0	1	1	0	0	1	1
Y/Z	0	1	0	1	0	1	0	1	0	1	0	1	0	1	0	1
XY/Z^2	0	0	0	0	0	0	1	1	0	1	1	0	0	1	0	1
XY/Z^2	0	0	0	0	0	1	0	1	0	0	1	1	0	1	1	0
X^3/Z^3	0	0	0	0	1	1	1	1	1	1	1	1	1	1	1	1
X^2Y/Z^3	0	0	0	0	0	0	1	1	0	1	0	1	0	1	1	0
X^2Y/Z^3	0	0	0	0	0	1	0	1	0	1	1	0	0	0	1	1
Y^3/Z^3	0	1	1	1	0	1	1	1	0	1	1	1	0	1	1	1
1	1	1	1	1	1	1	1	1	1	1	1	1	1	1	1	1
X	0	0	0	0	1	1	1	1	α	α	α	α	β	β	β	β
Y	0	1	α	β	0	1	α	β	0	1	α	β	0	1	α	β
Z	1	1	1	1	1	1	1	1	1	1	1	1	1	1	1	1

Figure 6.

Let, for instance, $q = 16$. Correspondingly the Hermite curve of degree $l = \sqrt{q} + 1 = 5$ has 65 points. The sequence of gaps in point P has the form

$$(0, - - -, 4, 5, - - 8, 9, 10, -).$$

Let us consider a code determined by divisor $G = 16P$. Since $W = 10P$ is a canonical divisor the code distance is $d \geqslant 8$ at $r \leqslant 11$. By analogy let us calculate the values for other values of d:

q	n	d	r	r'
16	64	8	11	15
		10	15	19
		12	16	23
		14	19	27
		16	21	31
		18	23	33
		20	25	37
		22	27	41
		24	29	45
		26	31	49
		28	33	53
		30	35	57

Given here for comparison is the value r' - the number of control symbols of a corresponding code on a rational curve. The table is presented as an example and illustrates the general property of codes on algebraic curves - given a big ratio N/g these codes behave much better than rational ones. If $q = 2^m$ good binary codes can be obtained from good q-adic ones by means of concatenation or transition to subfield over the subcode.

Let us consider, for instance, the simplest concatenation construction allowing the construction of binary analogues of codes cited in the table.

Let us put down each element of field F_{16} in the form of a 4-adic binary word using a certain basis of field F_{16} over F_2. Then we add yet another binary symbol equal to the sum of all the previous 4-adic symbols (general check of parity). As a result we get a binary code with a length equal to

$$n' = n(m+1) = 64 \cdot 5 = 320.$$

The number of information symbols is $k' = k \cdot 4$, so $r' = 5n - 4(n-r) = 64 + 4r$, while the code distance is $d' \geqslant 2d$ because in each block of 5 symbols there are at least two integers. We get the following binary codes:

q	n	d	r	r'
2	320	44	172	181
		48	180	199
		52	188	208
		56	196	226
		60	204	244
		64	212	262
		76	236	273

It should be noted that the notion of the Weierstrass point in the characteristic p has its peculiarities as compared to the characteristic 0. In the latter case all points, except the finite ones, have one and the same sequence of gaps $(1,2,...,g)$. The sequence of gaps $(N_1, N_2,..., N_g)$ is typical in the characteristic p and it is only the finite number of points that has another sequence of gaps (i.e. the Weierstrass points). In 1939 F.K. Schmidt indicated as an example the curve

$$Y^q + Y = X^m, \quad m > 2, \quad m \,|\, (q+1).$$

The genus of this curve is

$$g = \frac{(m-1)(q-1)}{2},$$

while the number of F_{q^2} points reaches the Hasse-Weil upper bound:

$$N = q^2 + 1 + 2gq = 1 + q(1 + m(q-1)).$$

The set of Weierstrass points on such a curve coincides with the set of all F_{q^2} points.

When $m = q + 1$, all points, with the exception of Weierstrass points, have a sequence of gaps

$$(1,2,...,q-1,q+1,q+2,...,2q-2,2q+1,...),$$

while the Weierstrass points have a sequence of gaps

$$(1,2,...,q-1,q+2,q+3,...,2q-1,2q+3,2q+4,...).$$

The Schmidt curves also lead to a row of interesting codes.

Codes on modular curves turn out to be particularly effective in instances of large lengths n. If the inequality

$$r \leqslant d + g - 1,$$

is put into the estimate

$$n \geqslant (\sqrt{q}-1)(g-1),$$

the resultant bound

$$r/n \leqslant d/n + \frac{1}{\sqrt{q}-1},$$

lies above the asymptotic Varshamov-Gilbert bound in a certain range of transmission speeds for sufficiently big values $q = p^{2m}$ (Yu.I. Manin, Some applications of algebraic geometry, Trudy MIAN, v. 168, Moscow, Nauka Publishers, 1984).

10. Covering Curves

In connection with codes on algebraic curves there arises the problem of constructing a curve of genus g over field F_q with a maximum possible number of F_q points. Here we shall outline an approach to solving this problem.

An irreducible curve passing through all the points of affine space A^n over F_q shall be called a covering curve and designated by the symbol $\Phi_{n,q}$.

As it was shown in the previous chapter for $n = 2$ each such curve has the form

$$\Phi_{2,q} = (X^q - XZ^{q-1})A(X,Y,Z) + (Y^q - YZ^{q-1})B(X,Y,Z).$$

In particular, when $A = X$, $B = Z$ we get an absolutely irreducible curve with parameters

$$g = \frac{q(q-1)}{2}, \quad N = q^2 + 1. \tag{4.20}$$

When $q = 2$ we have a binary elliptic curve with a maximum possible number of points $N = 5$:

$$Y^2 - Y = X^2(X - 1),$$

(here the Hasse-Weil upper bound is attained).

With $q = 3$ the curve $\Phi_{2,q}$ is a nonsingular quartic

$$X^4 - X^2Z^2 = Y(Y^2 - Z^2)Z,$$

having $g = 3$ and $N = 10$ (this, too, is a maximum curve but in this case the Hasse-Weil bound is too crude: $N \leqslant 14$).

Let us now consider another variant of selecting forms A, B:

$$\Phi_{2,q} = (X^q - XZ^{q-1})X - (Y^q - YZ^{q-1})Y.$$

In this case the form splits:

$$\Phi_{2,q} = (X^{q-1} - Y^{q+1}) - Z^{q-1}(X^2 - Y^2) =$$

$$= (X - Y)\left[\sum_{i=0}^{q} X^i Y^{(q-i)} - Z^{q-1}(X + Y) \right].$$

The remaining form

$$\sum_{i=0}^{q} X^i Y^{(q-i)} = (X + Y)Z^{q-1},$$

determines the nonsingular curve of C genus

$$g = (q - 1)(q - 2)/2, \quad \text{if char} = 2.$$

But if char $\neq 2$, then the polynomial $\sum_{i=0}^{q} X^i Y^{(q-i)}$ is divisible by $(X+Y)$ so we obtain an irreducible form of degree $(q-1)$:

$$Z^{q-1} = \sum_{i=0}^{q-1} (X^2)^i (Y^2)^{(\frac{q-1}{2}-i)}.$$

Let us find the number of points of the curves. The composite curve $\Phi_{2,q}$ passes through q^2 points at $Z = 1$ plus one point in infinity $(1,1,0)$ if char $= 2$. But if char $\neq 2$ then we should add yet another point $(-1,1,0)$. We must exclude from this number all the points of the straight line $X = Y$ not belonging to the curve C and also the straight line $X = -Y$. As a result we get the following expressions for the number of points of the curve

$$\left.\begin{array}{l} N = q^2 + 1 - q \\[2mm] g = \dfrac{(q-1)(q-2)}{2} \end{array}\right\} \quad \text{char} = 2,$$

$$\left.\begin{array}{l} N = q^2 - 2q + 1 \\[2mm] g = \dfrac{(q-2)(q-3)}{2} \end{array}\right\} \quad \text{char} \neq 2.$$

Now we shall consider covering curves of the smallest possible genus for A^3. Each such curve is an intersection of two surfaces:

$$(X^q - XZ^{q-1})A_1 + (Y^q - YZ^{q-1})A_2 + (U^q - UZ^{q-1})A_3 = 0,$$

$$(X^q - XZ^{q-1})B_1 + (Y^q - YZ^{q-1})B_2 + (U^q - UZ^{q-1})B_3 = 0.$$

Here A_i, B_j, $i, j = 1,2,3$ are forms from X, Y, Z, U.

These forms should be selected in such a way that the corresponding surfaces were irreducible while the curve at their intersection had the smallest possible genus.

Let us first consider $q = 2$.

The surfaces

$$\begin{cases} (X^2 - XZ)X + (Y^2 - YZ)Z = 0 \\ (Y^2 - YZ)Z + (U^2 - UZ)U = 0 \end{cases}$$

pass through the curve in the plane $U = X$. We shall add both expressions and make the substitution $U \to U' - X$. We get:

$$\begin{cases} (X^2 - XZ)X + (Y^2 - YZ)Z = 0 \\ U'(U'^2 + U'(Z + X) + X^2) = 0. \end{cases}$$

This curve is thus composed of the elliptic curve C' in the plane $U' = 0$ and the full intersection C of the cubic and the quadric. The composite curve has 9 points and the curve C has 7 points.

The genus of the curve, which is a full intersection, is defined by the formula

$$g = \pi - \delta,$$

$$\pi = \frac{1}{2} N_1 N_2 (N_1 + N_2 - 4) + 1.$$

Here N_1, N_2 are the degrees of the initial surfaces, π is the arithmetic genus while δ depends on the singularities of the curve.

The curve C has only one singular point $P = (0,1,0,0)$. This is an ordinary binary point (knot), so $\delta = 1$.

We have:

$$\pi = 4, \quad \delta = 1, \quad g = 3, \quad N = 7.$$

It is easy to see that the curve C has a plane nonsingular model of the type:

$$Y^2(U^2 + U + 1) + Y(U^2 + U + 1)^2 + U(U + 1) = 0.$$

Thereby we found a binary hyperelliptic curve of genus 3 passing through all the 7 points of the projective plane.

Now let us consider the curve by $q = 3$:

$$\begin{cases} (X^3 - XZ^2)X + (Y^3 - YZ^2)Z = 0 \\ (Y^3 - YZ^?)Z + (U^3 - UZ^2)U = 0. \end{cases}$$

After some transformations we obtain the curve

$$\begin{cases} (X^3 - XZ^2)X + (Y^3 - YZ^2)Z = 0 \\ Z^2 - X^2 - U^2 = 0 \end{cases}$$

in which $\pi = 9$, $\delta = 1$, so $g = 8$. It is easy to calculate the number of points of this curve. It is equal to $N = 13$. This is also a maximum curve. Here is a table of parameters of some covering curves.

q	g	N
2	1	5
2	3	7
3	3	10
3	8	13
4	3	13
4	6	17
5	3	16
7	10	36
8	21	57
8	28	65

11. Geometric Decoding

Whereas a code is determined by means of a projective model of a curve, the columns of the parity check matrix are points of a certain spatial curve of genus g and degree S which is normal in P^{r-1}. The length of code is evaluated as follows:

$$d \geqslant S - (2g - 2), \quad d = 2t + 2.$$

Let us define by t-chord the space stretched over t points of the curve. Not more than one t-chord passes through each point S of the space P^{r-1} (syndrome), this being equivalent to $d \geqslant 2t + 1$. Consequently, decoding is reduced to finding the points of intersection of the curve with a certain linear variety.

Decoding is performed in two stages:

1) Preliminary stage: to write down the equation of the curve's t-chord.
2) As the vector comes in the communication channel's output to calculate the syndrome S and find points of intersection of the t-chord passing through S

with the curve.

As it was shown in Chapter 1, in the event of a rational curve the matter is reduced to solve a system of linear equations. In the general case it is necessary to solve a system of polynomial equations by the method of elimination.

Comments

This chapter is based on the Serre book [14] (the geometrical part of this book). Moreover, certain chapters of Shafarevich [13], Shimura [16], and Lang [7] are used.

Two different constructions of the code on the algebraic curve are actually studied here. The first, more simple construction is connected with the projective model of the curve and identifies the code vector with the differential residues vector (§ 9). The second construction (generalized Jacobian codes) employs logarithmic differentials which naturally lead to the generalized Jacobian variety of a curve.

References

[1]   M. EICHLER, *Einführung in die Theorie der algebraischen Zahlen und Funktionin.* Basel-Stuttgart, 1963.

[2]   D.K. FADDEEV, *On the divisor class groups of some algebraic curves.* Dokl. V. 136, pp. 296-298 (1961).

[3]   W. FULTON, *Algebraic curves.* Math. Lecture Notes Ser., Benjamin, 1969.

[4]   R.W. HAMMING, *Error detecting and error correcting codes.* Bell Syst. Tech. J., 29 (1950), pp. 147-160.

[5]   A. HOCQUENHEM, *Codes correcteurs d'ereurs.* Chiffres 2 (1959), pp. 147-156.

[6]   Y. IHARA, *On congruence monodromy problems.* Lecture Notes, Univ. of Tokyo (1968-1969).

[7]   S. LANG, *Introduction to algebraic and abelian functions.* N.Y., 1982.

[8]   K. MAHLER, *Lectures on Diophantine approximations.* Univ. of Notre Dame. 1961.

[9]   B. SEGRE, *Lectures on modern geometry.* Edizioni Cremonese, Rome, 1961.

[10]   B. SEGRE, *Introduction to Galois geometry.* Atti Acad. Naz. Lincei Mem. Cl. Sci., Fis. Math. Natur. Ser. I (8), 8 (1967), pp. 133-236.

[11]   J.G. SEMPLE & L. ROTH, *Algebraic projective geometry.* Oxford, Clarendon Press, 1956.

[12]   J.G. SEMPLE & L. ROTH, *Introduction to algebraic geometry.* Oxford, 1949.

[13]   I.R. SHAFAREVICH, *Basic algebraic geometry.* Moscow, Nauka, 1972.

[14]   J.-P. SERRE, *Groups algebriques et corps de classes.* Hermann, Paris, 1959.

[15]   F. SEVERI, *Vorlesungen über algebraische geometrie.* Teubner, Leipzig, 1921.

[16]   G. SHIMURA, *Introduction to the arithmetic theory of automorphic functions.* Princeton, 1971.

Index